Conversations on the Beach

Studies in Environmental Anthropology and Ethnobiology

General Editor: **Roy Ellen**, FBA
Professor of Anthropology, University of Kent at Canterbury

Interest in environmental anthropology has grown steadily in recent years, reflecting national and international concern about the environment and developing research priorities. This major new international series, which continues a series first published by Harwood and Routledge, is a vehicle for publishing up-to-date monographs and edited works on particular issues, themes, places or peoples which focus on the interrelationship between society, culture and environment. Relevant areas include human ecology, the perception and representation of the environment, ethno-ecological knowledge, the human dimension of biodiversity conservation and the ethnography of environmental problems. While the underlying ethos of the series will be anthropological, the approach is interdisciplinary.

Volume 1
The Logic of Environmentalism
Anthropology, Ecology and Postcoloniality
Vassos Argyrou

Volume 2
Conversations on the Beach
Fishermen's Knowledge, Metaphor and Environmental Change in South India
Götz Hoeppe

Volume 3
Green Encounters
Shaping and Contesting Environmentalism in Rural Costa Rica
Luis A. Vivanco

Volume 4
Local Science vs. Global Science
Approacheds to Indigenous Knowledge in International Development
Edited by Paul Sillitoe

Volume 5
Sustainability and Communities of Place
Edited by Carl A. Maida

Conversations on the Beach

Fishermen's Knowledge, Metaphor and Environmental Change in South India

Götz Hoeppe

Berghahn Books
New York • Oxford

First published in 2007 by
Berghahn Books
www.berghahnbooks.com

Library of Congress Cataloging-in-Publication Data

A C.I.P. catalog record for this book is available from the Library of Congress.

British Library Cataloguing in Publication Data

A catalogue record for this book is available from the British Library

Printed in the United States on acid-free paper

ISBN-10: 1-84545-015-9 ISBN-13: 978-1-84545-015-1 hardback

Contents

Dedicated to the memory of Nediripil Haridasan Thulasi,
and to Korupattu Theyyan Krishnankutti,
with love and respect.

List of Figures, Maps and Tables

Figures

Maps

Tables

Preface

This book gives an interpretative account of a fisherfolk's environmental knowledge and their perception of human–environment relations. It attempts to identify the response of both of these to dramatic social, economic and environmental changes. I present ethnographic material as it has emerged from fourteen months of fieldwork in a South Indian fishing village and let this material guide me to understand the ways in which knowledge is expressed and its transformation debated. Particular attention is paid to the fishermen's everyday speech and their use of figurative language.

The focus of this study is on Chamakkala, a fishing village in central Kerala. Most of the fishermen in the village are Hindus and belong to the Araya caste. Until recently, many of the village's Muslims were involved in fish trading. I shall therefore try not to lose sight of this community, whose development has been closely related to that of the Arayas. Over the last decades, the fisheries of Chamakkala have been profoundly transformed in several ways. The sea itself has been affected due to the over-exploitation of marine resources and the operations of foreign fishing ships; new fishing techniques have been introduced and old ones abandoned; and the fish trade has been transformed profoundly by the opening of international markets for local catches and the new capability to handle bulk catches. Yet more profoundly, what the fisherfolk had hitherto perceived as innate to the environment, has been transformed by the action of other humans who are extraneous to their life-world. The village itself has become the target of state-induced efforts of development and, in addition, has experienced the manifold effects of the outmigration of workers. All these changes involve a community whose members were considered as polluting (because of their work of fishing) and, as such, unapproachable by high-status Hindus until not long ago. Inevitably, then, this book must also be about marginalisation and the struggle for upward social mobility.

I am convinced that the situation in Chamakkala resembles that of many places in the developing world. Marginalised at the fringe of a hierarchical society until recently, the fisherfolk are suddenly drawn into the struggle of 'modernisation' and 'globalisation'. For them, the present is a time of multiple uncertainties. In a situation like that, both their environmental knowledge and their relations to the environment are subjected to unanticipated tensions and transformations, but may also gain new importance for their orientational qualities.

The book is arranged as follows. In Chapter 1, I formulate the problems to be considered, develop a working definition of 'local knowledge' and address the conduct of my fieldwork and some of its methodological implications.

Chapter 2 introduces the setting of my fieldwork and the two social groups which have been most involved in local fishing: Hindu Arayas and Muslims. I review the migration of workers to the Arabian Gulf, which has affected the village's economy and local fishing profoundly. Finally, I describe the local conceptualisation of space as revealed by the usage of metaphors in everyday language.

Chapter 3 is concerned with fishing as a process of active knowledge making. Following a review of the technologies employed by Chamakkala fishermen during the last century, I give detailed accounts of two fishing trips, describe how fish are identified in the sea and how it is decided whether or not to catch them. It turns out that rather than following a fixed routine, the process of fishing is more like an improvisation or a 'performance' in which cognition is distributed among the team members. In its course, knowledge about the location of fish is generated and utilised. Fishing emerges as a social skill in two ways: with regard to the conduct of the team itself as well as with regard to the elicitation of knowledge from the observation of other boats. The latter forms a 'social environment' of a network of agents.

Chapter 4 presents ethnographic data on the fishermen's conceptualisation of the marine environment, from the sea's spatial structure and seasonal cycles to fish and their behaviour. While paying particular attention to figurative speech, I do not make prior hypotheses concerning the fishermen's ontology and the epistemological foundations of their environmental knowledge.

A main thread of the material of Chapter 4 is the fishermen's social relatedness with the sea. Chapter 5 elaborates on this topic by considering normative moral conduct as well as interpreting rituals that deal symbolically with the fisherfolk's relation to the sea. I continue by searching for an anthropological formulation for this relatedness.

Chapter 6 considers two state interventions that have directly affected the local practice of fishing, one by the colonial power and the other by the independent state. The first is a ban on the fishery of oil sardines, issued by the (British-ruled) Madras Presidency from 1943 to 1947; the second is the debate on the utility of a seasonal ban on trawling in (independent) Kerala State, raging on since the late 1970s. Both interventions have been regarded as (at least partly) misconceived by Chamakkala fishermen, albeit being of limited impact on the marine environment. Unlike these, the construction of dams in the Western Ghat mountains in the east is perceived to have affected the sea profoundly, challenging what has hitherto been regarded as its innate ('natural') character.

Chapter 7 begins by considering eclectic statements on environmental change, mixtures of local and apparently global knowledge. Locally, changes are commonly perceived in the context of the 'time of progress', which is contrasted to the 'old times'. In this epoch foreign knowledge has affected previously established conceptualisations in a number of distinguishable ways. A pertinent theme is the local response to imported foreign and powerful knowledge, a topic that cannot ignore the fisherfolk's enduring marginalisation in Kerala's society.

Acknowledgements

I have accumulated many debts in the course of the work that resulted in this book. I thank the people in Chamakkala and Chentrappini. I am most grateful to K.V. Vijayan and his family, namely his wife Susheela, his sons Sijeesh and Shinoj and his daughter Smitha, and I.B. Venu and his family, for their friendship and hospitality. Not only did Vijayan and Venu answer innumerable questions, but they also introduced me to many of my conversationalists and became friends who made my stay enjoyable. In addition I must single out K.T. Krishnankutti, E.V. Kumaran, K.V. Kunhiayyappan, K.K. Kunhiraman and K.V. Vasu for becoming my 'inner circle' of regular conversationalists. I thank all my other conversationalists as well.

I acknowledge the patient help and painstaking work which M.K. Sureshkumar did for me throughout my stay in the village. I also thank him for his interest, attention, ideas and humour. His optimistic response to many challenges ('this can be done') was always reassuring. Thank you, Suresh! I am very grateful to V.K. Ravindran and his wife ('teacher'), son Kannan and daughters Kanmoni and Surya. Thanks to Ravindran I was able to conduct my studies in the village, find my first house and make first contacts. I appreciate knowing him as an inspiring and knowledgeable scholar. If Ravindran was one of my 'patron saints' in the village, then K.K. Subramoniam was the other. He arranged for my contacts with the fishermen, gave me much support and advice and provided accommodation during my revisit in 2002. The many conversations with him (and the tasty food provided by Lalitha) made my stay enjoyable. While residing in 'Moscow' I much appreciated the company and friendship of Ambili and his family, Mammadukka, Mohammad, Jayarajan and the Thaththans. I am particularly grateful to the late N.H. Thulasi for cooking tasty food, washing my clothes and cleaning my small house. She and her family also cheered me up with their warmth and humour. I deeply regret her untimely death in December 2000. I thank my neighbour M.V. Kuttan and his family for their company while residing at the beach.

I am grateful to L.S. Rajagopalan (Thrissur) for his help, advice, inspiration and good humour. My visits to him belong to my most cherished memories of my time in Kerala. In Kunathangadi I am very grateful to Francis Kitangan, Mariyamman, Alex and Aaron for their friendship, kind hospitality and logistic support. I thank P.R.G. Mathur (Kozhalmangam) and his wife for their hospitality and inspiring conversations. I thank P.G. Kurup (Kochi) for explaining the oceanography of the *cākara* to me. At the Central Marine Fisheries Research Institute (Kochi) I thank N.G.K. Pillai and V. Chandrika. In Thiruvananthapuram I am grateful to V.R. Prabodhachandran Nayar for being a patient teacher of Malayalam and many things Keralese. I thank him and his family for their hospitality. Also in Thiruvananthapuram, I thank the Programme for Community Organisation (PCO), the South Indian Federation of Fishermen's Societies (SIFFS), the Kerala Census Office, the Kerala Swathanthra Matsyathozhilali Federation (KSMTF), the University of Kerala and the Centre for Development Studies (CDS) for the opportunity to use their libraries and archives. I am grateful to Jan Brouwer (Mysore) and his family for affiliating my study at CARIKS (Centre for Advanced Research on Indigenous Knowledge Systems), plentiful advice, cold Dutch beer and for cheering me up.

I acknowledge the scholarly advice and information provided by Caroline Osella (London), Filippo Osella (Brighton), David Shulman (Jerusalem), Shri Sivan Namboothiri (Berlin) and Francis Zimmermann (Paris). I am especially grateful to Gilles Tarabout (Paris). By introducing me to L.S. Rajagopalan he invoked a long chain of personal contacts that eventually led me to Chamakkala. I owe Cecilia Busby (Canterbury) special thanks for a careful reading of and extensive comments on an earlier draft.

I am grateful to Ute Luig and Helene Basu for supervising my thesis and to Nadine Sieveking and Jochen Seebode for critically reading parts of it. In the course of inspiring discussions Ute Luig made me see my ethnographic material in a new light. I thank Harry Falk, Michael Casimir and the late Georg Elwert for fascinating discussions. Rainer Kimmig and Axel M. Quetz helped me to deal with the font of diacriticals. Martina Weckauf and Bärbel Wehner scanned the photographs and Carmen Müllerthann processed the scans and reworked the line drawings. I thank Mark Stanton at Berghahn for his skillful and patient editorial work.

I am grateful to Carol Otto for her patience and skill in correcting the English in several drafts of the text. I thank her and Karin Brandmüller for help in dealing with the vicissitudes of bureaucracy. I am grateful for my parents' help and support.

Stefanie Märklin has endured much during the writing of this book. I thank her for her love, care, patience and understanding.

Financial support to carry out this work was provided by the NaFöG-Kommission of the Senate of Berlin and by the Deutsche Forschungsgemeinschaft (grant Lu-528/7–1). This essential help is much appreciated.

Note on Transliteration

All of the conversations with the people of Chamakkala were conducted in Malayalam, the language of Kerala. In this book, important Malayalam words are written in italics and have been transliterated according to the Library of Congress notation, a style followed, for example, by Tarabout (1986). In order to make the text more accessible, place names and personal names are not transcribed with diacritics. Unless otherwise noted (by the abbreviations pl., plural, and sing., singular) I write the plurals of Malayalam words by adding '-s' to the singular word. I do so because of the complications of the Malayalam plural construction.

Abbreviations

Arab.	Arabic
coll.	colloquial language
Eng.	English
fig.	figurative
inf.	infinitive
lit.	literal
Mal.	Malayalam
Skt.	Sanskrit
Tam.	Tamil

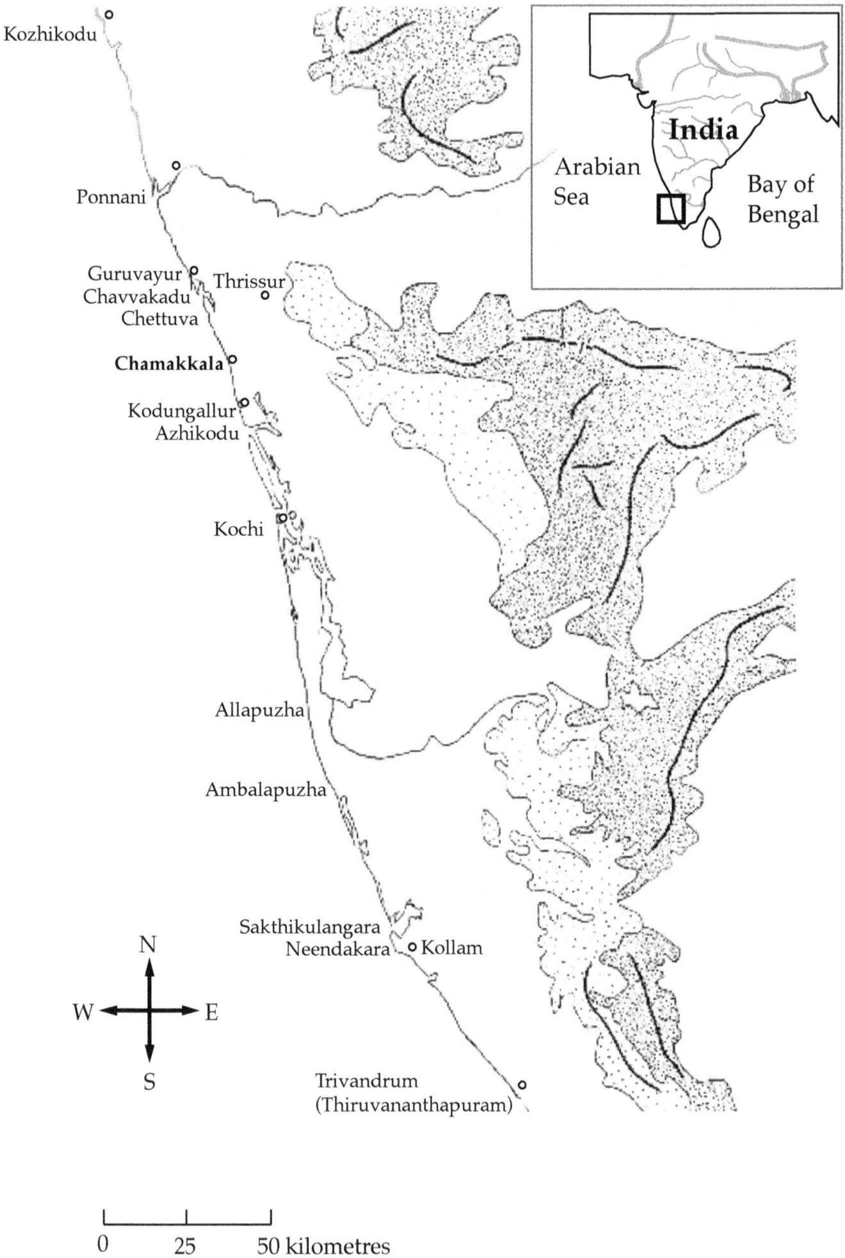

Map 1: Malabar coast.

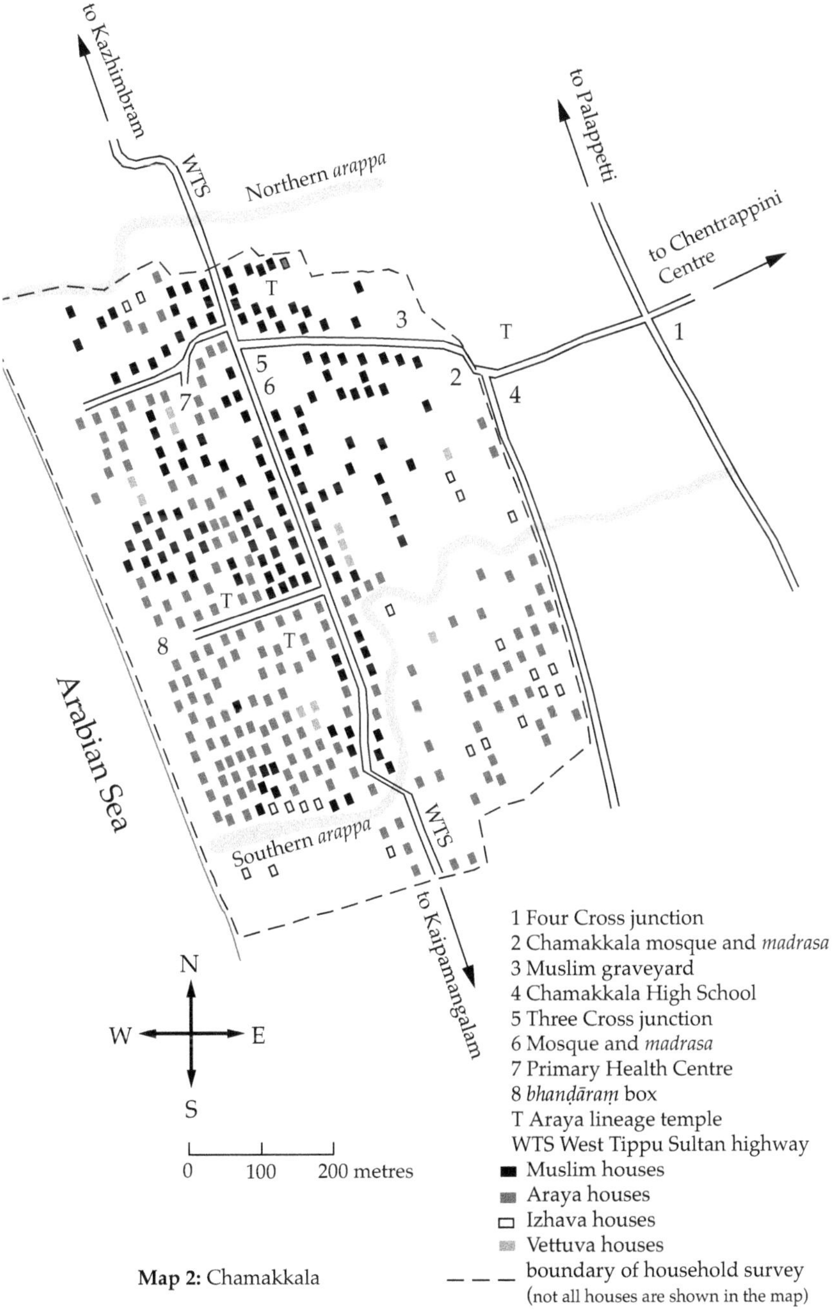

Map 2: Chamakkala

1 On the Beach with Krishnankutti

In November 1999, a few weeks after my arrival in the village, my co-worker Sureshkumar Muttichoor[1] (henceforth Suresh) and I were sitting on the beach in the village of Chamakkala with Krishnankutti Korupattu. Then in his late seventies, Krishnankutti had retired from fishing two decades ago. He had worked as a fisherman throughout his active life. Living nearby in a small house with his wife Leela, he was a daily visitor to the beach. His slim body, head bent forward, and his stiff, quick gait gave him a conspicuous appearance from a distance. He was fond of being on the beach. There would always be a reason to talk about some new and many old stories. Attracted by our talk, we were soon joined by Gangadharan Kochikkattu, a retired fisherman senior to Krishnankutti. Throughout their professional life, Krishnankutti and Gangadharan had been working on *vañci*-s, dugout boats propelled by rowing. Since the mid-1980s, these boats have been replaced by *vaḷḷaṃ*-s, large, motorised plank-built boats. We were also joined by Vijayan Kovil Tekku Valappil, forty-seven years old, an active fisherman on a *vaḷḷaṃ*. My questions induced a lively discussion about the village, its different communities and the changes that had taken place locally in the last three decades. Krishnankutti and Gangadharan kept comparing the 'old times' (pl. *paḻaya kālaṅṅaḷ*) with the present 'time of progress' (*purōgamiccụ kālaṃ*; lit. 'progressed time'), a distinction I was to hear often during my fieldwork. Krishnankutti agreed that there had been immense changes on the land. There was the new road to the beach, new houses and the money that had been sent home by men who worked as migrants in countries in the Arabian Gulf. But even more dramatic changes, he lamented, could be seen in the sea.

Before considering what Krishnankutti and Gangadharan said, it will be necessary to locate Chamakkala in the geography of Kerala. The village is situated on the shores of the Arabian Sea. From here, a flat alluvial plain extends about fifty or sixty kilometres inland (i.e. towards the east), where the terrain begins to rise towards the Western Ghats, a chain of mountains running parallel to the south-west coast of India. Reaching an elevation of more than two thousand metres, the Western Ghats form South India's principal watershed (see Map 1). All except three of Kerala's forty-four rivers flow westward and (used to) carry abundant water into the Arabian Sea during the south-west monsoon (June to August). Since the late 1950s, a number of dams have been built to use the river water for irrigation, for drinking and to generate electricity.

Gangadharan Kochikkattu: In the old days porpoises (*ēti*), big shark (*sravụ*) and stingray (*tiraṇṭi*) tore our nets!

Krishnankutti Korupattu: We used to catch stingrays (*tiraṇṭi*)!

Gangadharan: We caught the small ones, but the big ones were escaping and by doing so they tore the nets.

Krishnankutti: At that time, bigger nets [i.e. thicker threads] were used. The net was made of *vakkụ* [a small bush with strong fibres]. The thread of it would have such a size. (He shows his index finger.) Using that we managed to catch sharks of about two thousand kilograms and more.

Gangadharan: The thread was of this thickness. (He also shows his index finger.)

Krishnankutti: Shark is still there, but they can't catch it now.

Vijayan Kovil Tekku Valappil: It is not coming to the shore from the outer sea (*puṟaṃ kaṭal*) any more.

Krishnankutti: It can't come now. Because of these ships and boats. They are all over and move around in the sea. In that time the noise (*śabdaṃ*) of the [ship] engines was not heard in the sea.

Gangadharan: In that time shark was following the shoals (sing. *pelappụ*) of oil sardines (*cāla*). It was playing with them and ate them. Then on the sight of a shoal of oil sardines we would follow the shoal of the oil sardines with the nets drawn. Then the sharks might be coming to eat the oil sardines, and thus we catch the shark. But the sharks will try to jump over the net in order to escape from it.

Krishnankutti: We tried to keep them in the net.

Vijayan: We can see that the sharks try to escape from the nets by crawling out and jumping.

Gangadharan: First it jumps within the net and then it attempts to dart into the sea to escape. But the net is strong.

Krishnankutti: On attempting to escape [by moving through the net] sometime the necks of the sharks will burst within the net and in other occasions the belly of the fish breaks when the net is tightened. But at any cost the rope won't be breaking. Then because of us tightening the ropes the sharks die (inf. *marikkuka*) in the net.

Gangadharan: The size of the thread should be of this size. (He shows his index finger.) If the fish [he means: shark] is not dead then we kill (inf. *kolluka*) it by stabbing a knife into their throat. Then it is pulled into the boat.

Krishnankutti: On some occasions we got twenty to twenty-two sharks in one net. I mean in one catch (*peytu̱*).

Gangadharan: Nowadays it is not seen. I believe that it is in the outer sea (*puṟaṃ kaṭal*).

Götz Hoeppe: Are there fewer fish in the sea now?

Krishnankutti: At that time, the small fishes like, I mean, almost all species (sing. *ellataraṃ*) of fish, such as shark (*sravu̱*), mackerel (*ayila*), oil sardine (*cāla*), prawns (*cemmīn*), big-jawed jumper (*parava*), jew fishes (*kora*), white sardines (*cūta*), soles (*mantal*), silver bellies (*mullan*), anchovies (*manaṅṅu̱*) and so on were available. Now there are less [many fish].

Suresh Muttichoor: Why are there less many fish now?

Krishnankutti: Mainly it is because the [trawling] boats and [fishing] ships always move around in the sea. The ships move around even in the region where the fish reside and hatch (*mīn parkkuna podiṅṅu̱ undakuna*). The ships move to the residence (*parkkuna stalaṃ*) of the fish and collect [i.e. catch] them from there. The ships and boats now move even to the outer sea (*puṟaṃ kaṭal*) for fishing. Because of this the fish withdraw backwards (coll. *a enna ekattine mīn kayaṟi kayaṟi nilkkukayānu̱*). Fish move even farther behind, but they are not lost (*mīn illyate avukayilla*).

Gangadharan: The reproduction is destroyed (*ulpādunuṃ naśiccu̱ pōkuṃ*).

Krishnankutti: Yes, the reproduction is destroyed. For us [to benefit us] fish has to arrive near the seashore. The arrival of fish has become rare. The fishes have climbed (inf. *kayaṟuka*) to the deep sea (*āḻa kaṭal*). The fishes have moved to the silent deep sea.

Krishnankutti: The sea of that time and of this time differs a lot. In that time so many kinds of fish were in the sea. Nowadays they are not seen. These fish do not come to the shore (*kara*) any more. (He exclaims) In that time!

Gangadharan: How can you say that the sea has changed?

Krishnankutti: We can say that the sea has changed (*kaṭalu̱ māri*). I mean the constitution of the sea itself (*kaṭalinṟe prakṛti tannu̱*) is entirely different now.

Suresh: Can you explain the difference?

Krishnankutti: In that time the eastern mountain water (*mala veḷḷaṃ*) came to the sea and the sea was shaken (inf. *iḷakkuka*) a lot by it. I mean, when the mountain water moves towards the west and the sea waves move towards the east then they will thoroughly agitate (*śarikku̱ iḷakkuṃ*) each other in the sea. The sea will be agitated (*kaṭalu̱ iḷakīttu̱*).

Vijayan: In that time the sea was agitated (*iḷaccu*) a lot. The sea became rough, ten times more than nowadays. A small wave of those days would be enough to kick us [present-day fishermen]. But the old ones [fishermen] resisted these waves. If an old man would be pushed by a wave, then people like me would be pushed ten times further. I say this in order to give you an idea of how strong they used to be then.

Krishnankutti: [In the old time the] agitation of the sea (*kaṭalu iḷakkaṃ*) was ten times stronger [than today]. Today, even when the sea is calm (*kaṭalu irikkuna*, lit. 'sitting down') these people do not go to work [because of the alleged roughness]. But for us that was nothing, we used to work even in the roughest [sea]. In that time fishing in the rough sea was thrilling to us. Even in the rainy season we would be sweating by the work [of fishing].

Suresh: Is the sea calm now?

Krishnankutti: The sea cannot be made as calm (*śāntaṃ*) now as it was then. In that time there was this shaking (*iḷakkaṃ*) and then the sea was cooled down (*tannuccu*).

Gangadharan: It is said that by the arrival of the mountain water the shaking (*iḷakkaṃ*) of the sea is stopped. After the agitation (*kṣōbhaṃ*) the sea used to be flat (*marggaṃ*). But now the rivers are full of dams (*anakettu*). Rainwater does not enter the sea any more and the agitation (*kṣōbhaṃ*) of the sea does not come to a halt (*nilkkunilla*).

Krishnankutti: After the mountain water agitation the sea was not getting rough in those days. But now it is different. Even a small wind can make the sea rough. Today the agitation of the sea is not coming to a halt (*kaṭalinṟe kṣōbhaṃ nilkkunilla*). That is only happening by the arrival of the mountain water. But the mountain water will not come any more because of the dams. Today the mountain water does not arrive any more. The water from the dams is used for drinking.

Götz: Some people say that in the olden days you could just go into the sea and wait for the fish to come.

Krishnankutti: That is loose talk of the people.

Gangadharan: We would search for it [the fish]. If it is not seen then we come back.

Krishnankutti: On some occasions we would move all around for fish but we won't see anything. Thus we would return. But in that way we would get a good catch [eventually]. I mean, if a certain point [in the sea] was empty on our [first] visit, then there may be fish at the second time. This happens sometimes. And you have to be aware that this is not a pond (*kuḷaṃ*). This is the sea.

Vijayan: More than that: the fish is a living matter (*jīvanulla vastu*). It is always moving.

This conversation gives a panoramic overview of the main topics considered in this book. In talking about the old times (pl. *paḻaya kālaṅṅaḷ*), Krishnankutti and Gangadharan are excited. Referring to the past they keep switching between present tense and past tense. One may guess how much their memories are alive, being relived during the conversation. Yet so much has changed between then and the present 'time of progress' (*purōgamiccu kālaṃ*). Both old men talk about the violence of fishing, the cruelty of killing sharks by cutting their throats with a knife, yet they also talk about the strength and stamina needed to be able to do so: a strength that most present-day fishermen lack, as Vijayan, a middle-aged fisherman, readily concedes. Today's fishermen are said to have weak bodies and rely on machines. The three men also talk about the decline in the fish population, about decreasing catches and missing species. They talk about trawling boats and fishing ships disturbing the sea and the fish moving away to the 'outer sea', and about the present lack of sharks and stingrays. And they talk about the difficulty, or, rather, impossibility, to predict the movements of fish.

In considering the constitution (*prakṛti*) of the sea (*kaṭal*), Krishnankutti and Gangadharan distinguish between two states. One is a state of agitation (*kṣōbhaṃ*) or shaking (*iḷakkaṃ*), which they link to the south-west monsoon (*kāla varṣaṃ*). Krishnankutti explains the agitation as a collision of (monsoonal) waves moving from west to east and the rainwater from the mountains (*mala veḷḷaṃ*), which comes to the sea through the rivers and moves from east to west. In its other state, the sea is calm (*śāntaṃ*), cool (*tanuppu*) and flat (*marggaṃ*). The adjectives describing either state are similarly used in discourses about the emotion and constitution of human persons and bodies. Lacking the cooling due to the mountain water, Krishnankutti laments, the sea is in a permanent state of agitation nowadays: it may become turbulent at any time. Formerly, he claims, it was agitated and cooled down profoundly thanks to the mountain water arriving from the east. But, since dams have been built in the mountains, the rainwater is lacking and this process of cooling is disrupted. Thus, the sea and the land are related to each other. Or, more precisely, the sea depends on the mountains to be cooled. As I was to hear later on, the fishermen wish the sea to cool down because the fish are said to prefer staying in cool environments and would then be, therefore, attracted to the coast.

In these early days of fieldwork I did not realise that the agitation of the sea by the mountain water is linked to a popular myth. A few months later, in July 2000, after the south-west monsoon had begun, Krishnankutti (Figure 1.1) complained that the mountain water was lacking due to being stored in the dams. Then he continued to explain the myth:

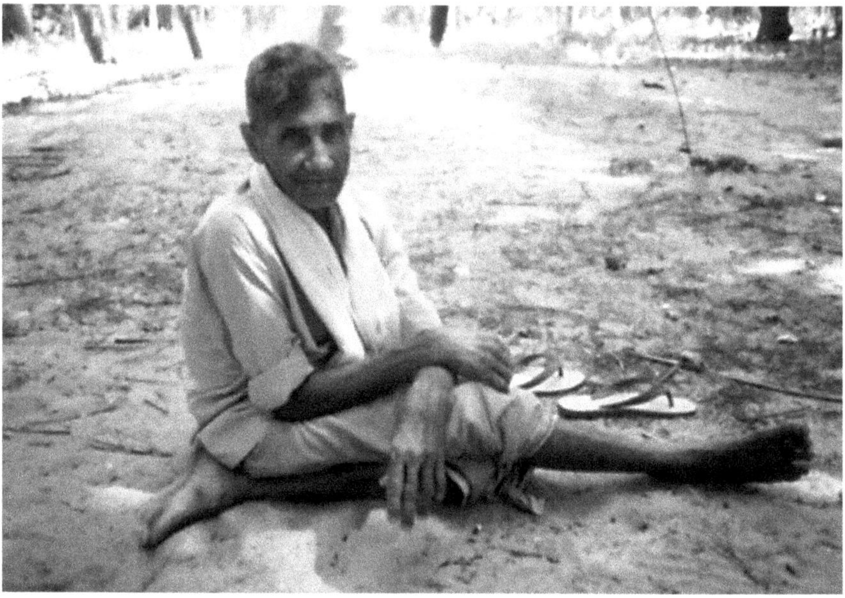

Figure 1.1: Krishnankutti Koruppatu

Krishnankutti: It is said like that: The sea is a woman (*strī*). There was a sort of promise (*vāgdattaṃ*) between the sea and the mountains [to meet intimately]. If the sea rises from here to there [mountains] then the entire world (*lōkaṃ*) will be destroyed. 'I shall come there [to the sea]' is a promise, which the mountains gave: 'I shall come through the rivers'. That is the mountain water (*mala veḷḷaṃ*). It has to come. Only then will the anger (*kōpaṃ*) of the sea be calmed down (*śantata uṇḍāvūkayolluḷ*). It was like that.

Götz: Why does the sea want to come to the mountains?

Krishnankutti: Why the sea wants to go to the mountains? The mountains are a man (*puruśan*). This here [the sea] is the woman (*strī*). Then at least once a year they have to unite (inf. *yōjikkuka*). But if the sea goes to that side [of the mountains], if it climbs up to the mountains, then there would be no world (*lōkaṃ*). Only water. So they made the promise that this will never happen. 'I will come to that side' [to the sea; say the mountains]. At least once a year it comes through the rivers. Isn't it like that? This water?

The cooling of the sea at the onset of the south-west monsoon, apparently so important for many fishes' seasonal arrival near the beach, is here linked to a cosmological process of the sexual union between the sea and the mountains.[2] As mentioned in the first conversation, the state of the sea is related to that of the land: it depends on the cooling effect of the mountain water. Unlike Krishnankutti's earlier explanation, the myth emphasises that, as subjects, sea and mountains are capable of moral judgement and intentional action. In arranging their intimate meeting they consciously agree to protect the population of Kerala. A parallel of the two viewpoints is that they assume the arrival of the mountain water, or the intercourse between the sea and the mountains to happen regularly. As mentioned by Krishnankutti in the first conversation, this rhythm has been disrupted by the construction of dams in the mountains. These dams store the mountain water and prevent it from entering the sea forcefully. As a consequence the sea cannot cool down any more. It stays hot permanently.

But who has built the dams? Like other men, Krishnankutti explains that the *kiḻakkaranmar*, the 'people of the east', have done so, and he surmises (correctly) that the state government has been involved in their construction as well. Of course, that these dams impede the mountain water is a negative effect. But Krishnankutti also knows that the dams provide drinking water and electricity. Since his own little house was supplied with electricity eight years ago, he has come to appreciate it, even though he uses his supply only to run two feeble light bulbs. (He also knows that the eastern, inland parts of the village had been supplied with electricity more than a decade earlier.) Unable to classify the dams as either 'bad' or 'good', Krishnankutti is aware that the 'time of progress' has come with an ambivalent effect upon local people.

In early September 2000, towards the end of the south-west monsoon, Suresh and I had another conversation with Krishnankutti about the mountain water. A few days earlier, the annual ban prohibiting bottom trawling and purse-seine fishing had been lifted. Introduced by Kerala's state government in 1988, the ban is meant to help the fish stocks regenerate during the monsoonal spawning season and benefit artisanal fishermen. Bottom trawling and purse-seining are not employed by artisanal fishermen in Kerala, but have been introduced by capitalist fish merchants. After the ban ended, trawlers and purse-seiners were seen again from the beach in Chamakkala. Krishnankutti has an ambivalent opinion about their effect upon the sea and the fishes:

Krishnankutti: When the fishes move to the west we say that they will ascend and ascend (inf. *kayaṟuka*). Now [in early September] they will come descending from there (coll. *iraṅṅi iraṅṅi varuṃ*). The ships [i.e. trawlers and purse-seine boats] are the reason. The ships move to the breeding place (*mīn pārkkuna dikil*, lit. 'fish [egg] laying direction'). When the ships move in the direction of the place for laying eggs, the fish move to the east (*kiḻakkụ*) from the west (*patiññāṟụ*). The big mackerel (*valiya ayila*) comes because of that.

Götz: People say that the fish come because of the noise (*śabdaṃ*) of the ships, don't they?

Krishnankutti: The ships make waves while they move. The water is agitated (*iḷaccu*) and boiled (inf. *tilaccu mariyuka*) because of that.

Götz: The sound is no problem?

Krishnankutti: The sound is not the main problem. The water is agitated (inf. *mariyuka*, lit. tumbling) because of the movement of the ships ... Because of the ships the fishermen from here have problems (sing. *dōṣaṃ*). When the ships started to operate the price for fish went down.

Like the verbs used to describe the sea's states of agitation and calmness, the verb *tilaccu mariyuka* refers both to boiling as a thermometric state of hotness and to the emotional condition of 'overwhelming with rage' or being 'highly excited and haughty' (Warrier et al. 1999: 474). In local, colloquial usage it has almost exclusively the former meaning. In either case, because of the ships' movement in the (western) sea, the fishes are moving to the shore. As such, while in the past the fish used to come to the shore during the monsoon because of the cooling of the nearshore waters, now the fish are driven to the shore after the monsoon because of the agitation of the water by the ships' engines. Thus, the combination of the construction of the dams in the east and the introduction of the trawling ban has introduced a new kind of regularity: a regularity that is not understood to be due to a cosmological rhythm of intercourse between the sea and the mountains as intentional beings. Rather, it is due to the obscure actions of the 'people of the east' (who build dams), the state government (which introduces a trawling ban), capitalist fish merchants, politicians and even movie actors (who own trawlers and purse-seine boats). Even though not mentioned by Krishnankutti, the influence of foreign, international fishing vessels is yet another disturbing influence that is widely discussed among local fishermen.

Altogether, these three conversations hint at how much the sea is embedded in a relationship with the land and with the people. Like human beings it can become heated and cool, agitated and peaceful. The building of the dams and the effects of the ships show that humans are powerful enough to have an impact on the marine environment. But it is not the fishermen who exert this influence. The dams in the east have been built by the 'people of the east' and the state government. In the west, i.e. in the sea, fishing is affected by privately owned ships, some of an obscure international origin, and regulated to at least some extend by the state and national governments. Thus, the world of the fishermen is encroached upon from both sides by forces that are difficult to influence and often remain obscure.

The Fisherfolk and the Kerala Model

The fisheries in Chamakkala have become thoroughly embedded in contexts that extend far beyond the village and the coastal area. Decisions of profound influence on local fishing and the village are made in the state capital Thiruvananthapuram and in the national capital New Delhi. As Chamakkala fishermen have learned from newspapers, the activity of foreign fishing vessels in India's Exclusive Economic Zone (which extends about 370 kilometres into the sea), has been enforced by the International Monetary Fund (IMF) and the World Bank. Both are based in Washington, DC (USA). The price of prawns, caught by Chamakkala's fishermen for export, is determined in markets in Korea, Japan, the European Union and the USA. Last but not least, part of the boats and equipment used at the beach were bought with money earned by migrant labour in countries in the Arabian Gulf.

I have wondered whether, by studying the local impact of these many, and often major, changes, one could learn something for the future of other places in the developing world. Kerala seems to be unique in a promising way. For many years, it has been a 'favourite anomaly' of development economists (Wallich 1995: 19). While the average per capita income is low, life expectancy is high, literacy is approaching 100 percent and the public health system is well developed (Parayil 2000: 1–2). These are the characteristics of the 'Kerala model of development', which some development experts wish to transfer to other regions of India and the developing world (Jeffrey 1992; Parayil 2000). However, one must not forget that a unique background made this development possible. Not only is its religious demography unique (approximately 60 per cent Hindus, 20 per cent Muslims and 20 per cent Christians; see Tarabout 1997b: 246), but there has been a long history of political mobilisation in the state, dating back at least to the 1920s (Menon 1994). In 1956, Kerala was the first state worldwide where a communist government was elected democratically.

Doubt as to whether the transfer of the 'Kerala model' to other regions is possible or even desirable has been growing. Certainly, it does have its dark sides. The suicide rate in Kerala is three times the national average, and, while the unemployment rate amounted to 25 per cent officially in the mid-1990s, it was probably closer to 35 per cent (Wallich 1995: 19). The situation of the fisherfolk, and of fishing in Kerala more generally, is another dark side of the 'Kerala model'. Statistical data show that the fisherfolk, for a long time marginalised at the fringe of a predominantly agrarian society, lays behind in terms of landholdings, health conditions, child mortality and literacy. It has remained an 'outlier' of the 'Kerala model' of development (Kurien 2000a).

The considerable importance of outmigration and the influx of remittances has raised doubts about the sustainability of Kerala's economic development (see Parayil 2000). The importance of foreign remittances seems particularly precarious in the light of fluctuating politics in the Arabian Gulf states. In the long run, the number of migrants from India to this region peaked in the 1980s, but declined sharply thereafter, especially since the Gulf war of 1991. Since then, labour migrants from other parts of India and other Asian countries have become serious competition to migrants from Kerala.

The fisheries point to some further problems of Kerala's development: the over-exploitation of natural resources and environmental degradation. As I shall elaborate in Chapter 6, these problems are thoroughly enmeshed with the development of an export-oriented fisheries since the Second World War.

Leading Questions

During the course of my project three groups of major questions have emerged. These are:

1. How do the fishermen conceive of the marine environment and their relation to it? In what terms do they communicate about it?
2. How (if at all) do the skill and improvisation needed in daily fishing operations interact with a relatively structured body of environmental knowledge?
3. How do the experience of the outside world and the changes seen in the local world, affect human–environment relations? How does the fishermen's knowledge fare in confronting the state, colonial as well as independent?

Locating 'Local Environmental Knowledge'

Let me consider the words making up the term 'local environmental knowledge'.[3] The immediate allusion of the 'local' is my fieldwork location, the village Chamakkala and its adjacent villages in the particular coastal strip, as well as the sea nearby. But, rather than considering it a static space, I follow Arjun Appadurai in considering locality as 'primarily relational and contextual' (1996: 178). Indeed, the world of the village is constantly produced and reproduced, relations being created, sustained or abandoned. As everywhere, global connections play an increasing role in these processes. Environment is a relational term as well, for there could be no environment without the environed. More than locality, environment is relational in a spatial sense, and I shall understand it in the sense typically associated with ecology (Wagner 1977, 1981; Ingold 2000). By knowledge I refer to what is known about the environment both in a conceptual, epistemic way and in contexts of practical engagement, especially in fishing. I refer to what the fishermen know as knowledge and to how they know it as epistemology. Local environmental knowledge, then, is the knowledge necessary for producing and reproducing a locality and an environment.

Locality and environment are likely to be composites of ecological, technological, biological and social realms, which cannot be clearly distinguished from one another. Likewise, the fishermen's knowledge of these domains must be equally composite. I am prepared that the categories 'ecological', 'technological', 'social' and so on 'only weakly convey the structure of agricultural knowledge in a place like India, where these domains are not so separated and where knowledge of water, animals, soil, persons, deities, crops, markets, and manure form part of an interacting whole' (Appadurai 1989: 174).

Inevitably then, knowledge about the environment must be knowledge about the social environment as well. I shall show that such interrelations between apparently distinct categories are characteristic for the knowledge of Kerala fishermen, which is further complicated by its uneven distribution and its idiosyncratic, often unsystematic, character. One may note that, in comparing or blending knowledges of entities that are as distinct as, say, deities and markets, one is, in fact, dealing with different ontologies. These different ontologies suggest the usage of similarly different epistemologies, i.e. ways of knowing. They demand different practices in the generation and transmission of knowledge. In addition, does not any delineation of knowledge result from a specific, and highly individualistic, communicative process between local informants and the anthropologist, as Valentine Daniel (1984: 47) puts it? To the ethnographer who sets out to explore a people's knowledge, a serious headache seems guaranteed. Too easily she or he may construct a 'knowledge' or even a 'knowledge system' that either is an idiosyncratic construct or may not even represent anything meaningful for the natives themselves (see Bloch 1992). As Roy Wagner writes in his book *The Invention of Culture*, 'the anthropologists invents 'a culture' for the people, and *they* invent "culture" for him' (1981: 11). Am I not, similarly, making a knowledge for the people? Do they make a knowledge for me?

This possibility is one reason to be wary of the notion of a 'knowledge system'. Another reason is, as Paul Richards has noticed, that a fixed, formalised 'system' of knowledge would be a hindrance for agricultural work; since being a farmer necessitates reacting quickly to unexpected, changing circumstances, a fixed knowledge system would impede the performance of farming (1993: 62). Lazare Séhouéto's (1995) work points in the same direction and illustrates the flexibility of small-scale agriculturalists, who skilfully accumulate knowledge through experimentation, often yielding particularistic solutions that are not, and cannot be, systematised formally. Much the same is to be expected in regard to the fisheries.

Anthropologists studying knowledge have concentrated on its bodies of assertions and the processes of its transmission in a culture (see Sperber and Wilson 1995; Barth 2002), whether they focus on the 'collective representations' shared by the members of a social group (as in French and British social anthropology) or concentrate on the individual mind in cultural perception (as in American cultural anthropology). In recent years, several studies have emphasised the embeddedness of knowledge in human relations with their environments, some following a relational approach (see Descola and Pálsson 1996; Ingold 2000). As in hunting, the aim of fishing is to kill life, and thus the fishermen's relation to the environment is fundamentally (and continually) enmeshed in acts of violence (see Valeri 2000). Issues of morality must be involved.

Apart from such issues that bring human-environment relations close to themes which are central to human identity, I agree with Maurice Bloch that an epistemologically reasonable account of environmental knowledge can only be obtained from 'understanding the combination of (1) constraints coming from the world as it is, and as it presents itself as an opportunity for human production, together with (2) the particular cultural history of groups and individuals, and (3) the nature of human psychology' (1998b: 43). By alluding to human psychology, to the 'internal' mind, he refers to studies by developmental and cognitive psychologists, linguists and anthropologists on the role of innate cognitive structures in conceptual development (e. g. Keil 1989; Hirschfeld and Gelman 1994; Astuti et al. 2004). Such works now yield often surprising insights into classical ethnographic matters such as ethnobiological classification (Atran 1990, 1998; but see Ellen 1993), anthropomorphism (Boyer 1996) and the choice of animals and plants as ritual objects (Bloch 1998b). They have also led to yet another flaring up of the debate over nature versus nurture (see Whitehead 2001).

Constraints placed by innate cognitive structures may shape the constitution of environmental knowledge in important ways, and I shall draw attention to where I perceive them as relevant in my fieldwork data. Yet as an ethnographer I have to stick to what is available to ethnographic study. I cannot but study what is 'external' to the mind, just because 'culture and social institutions are external, interactive, processual, historical realities', as Alfred Gell (1998: 127) reminds one. He insists that the specific subject-matter of anthropology is 'social relationships' and that anthropological theories must, therefore, be 'theories about social relationships' (ibid.: 5). My aim is to produce an account of the fishermen's knowledge that is recognisably anthropological in this sense.

In Kerala's Malayalam language, the verb *aṟiyuka* means 'to know'. The appertaining noun is *aṟivu̧*, 'knowledge'. While Chamakkala fishermen commonly use the verb, they rarely make use of the noun. Rather, if they talk about a specific known entity or process, they may refer to the 'fisherman's calculation' (*matsyatoḻilaḷuṭe kanakku̧*) or a certain rule (*niyamaṃ* or *citta*). A rule is often a specific memorised fact, for example that 'the south-west monsoon begins on the fifteenth day of the *iṭavaṃ* month [of the Kerala calendar]' or that 'oil sardines are caught after the tenth day of the *karkkaṭakaṃ* month'.[4] When reference is made to the fisherman's calculation, a mixture of specifically known facts, memories of previous experiences and practical skill is usually alluded to. For example, the incentive to go fishing in the sea on a particular morning may be derived from one's calculation (i.e. estimation) of the marine and market conditions at that time. Other knowledge may be in the form of a legend (*eitihyaṃ*) or a story (*kata*). The intimate encounter between the sea and the mountains, mentioned in the conversation at the beginning of this chapter, is an example thereof. In other cases knowledge may not belong to a category that is explicitly named. Whether 'rule', 'calculation', 'legend' or 'story'[5]: knowledge of this kind is distinguished from, if not contrasted to, *śāstraṃ*, 'scriptural knowledge'. The latter is systematic, formalised knowledge, often found in a religious context or in astrology and codified in texts. The Vedas are *śāstraṃ* par excellence. Reference to *śāstraṃ* is often made in a fairly generic way. Usually, *śāstraṃ* is considered as superior, more powerful and more universally valid than 'rules', 'calculations' or 'legends', which are often regarded as particularistic, i.e. applicable only in specific circumstances. In another sense, this perception may be based on the association of *śāstraṃ* with Brahmin dominance in local discourse, at least as far as the reference to scriptural traditions is concerned. As I shall elaborate in this book, in Chamakkala the allegedly Western 'fisheries science' has immediately attained a status similar to that of *śāstraṃ*. Thus, the vocabulary pertaining to knowledge refers implicitly to different kinds of knowing and hints at the possible importance of relations of power in discourse. On the other hand, fishermen in Chamakkala concede that there are matters which cannot be grasped by humans at all, or perhaps only in part. For instance, I was reminded that the exact places of fish in the sea and the changes of the weather and the currents in the sea cannot be predicted by anyone.

Knowledge, Language, Conversations

As the title of this book suggests, I consider conversations as central to the process of my fieldwork. The cooperation with my co-worker Sureshkumar Mutticho0r reminds me of the view taken by Stephen Gudeman and Alberto Rivera (1990). I agree with them that 'fieldwork is a perpetual discussion' and I am sympathetic to their proposal of 'the model of conversation as the practice of anthropology and as the activity of other cultures' (Gudeman and Rivera 1990: 4). This notion resembles that of Valentine Daniel who termed the conversation between fieldworker and informant the 'locus of the cultural process' (1984: 43). Daniel quotes Claude Lévi-Strauss as going as far as to equate the craft of anthropology with this conversation: 'For anthropology, which is a conversation of man with man, everything is symbol and sign, when it acts as intermediary between two subjects' (Lévi-Strauss 1966: 115). One may say that the conversation between the 'researcher' and her or his 'informants' includes language. But it is not restricted to it. Much rather, argues Daniel, the dialogic aspect of culture includes many non-verbal signs as well (1996: 198).

Nevertheless, if my fieldwork was to focus on conversations, language had to be crucial. Thus, before leaving for the field, I was struck by Maurice Bloch's claim, drawn from recent progress in the cognitive sciences, that knowledge is largely non-linguistic and can be expressed in language only at the price of seriously distorting it (1991, 1992, 1998a). Bloch's insistence that attentive and skilled participant observation may reveal much non-linguistic knowledge is well taken, but altogether I have grown confident that attention to language, in a symbiosis with participant observation, can successfully address many of the shortcomings mentioned by Bloch.

It turns out that Bloch's 1992 paper, in which he sets out to demonstrate the importance of non-linguistic knowledge, makes use of language at a central point. He argues that central concepts of Zafimaniry society (Madagascar) can be described as a series of intertwined 'mental models'[6] which are not language-like. A persistent theme in this conceptualisation is the idea of maturation as a process of hardening, regardless of whether stages in the life cycle, a marriage or a house are concerned. The prototypical exemplar of hardness is *teza*, the core of the trunks of certain trees (Bloch 1992: 140). In conceptualising the maturation of a house, not only the material *teza*-s used in building the house are alluded to, but also *teza*-ness as a metaphor for matured people and established houses (ibid.: 141). Thus, 'hardening' is at least in part understood figuratively and can be expressed in language, as Bloch himself demonstrates in his essay. This point is supported in a more general way by Hill and Mannheim (1992: 382), who note that it is problematic to separate 'language' and 'nonlanguage' and then 'relate' one to the other. I subscribe to their viewpoint that 'language, culture, and meaning have inextricably contaminated each other in the course of doing ethnography' (ibid.: 382–383).

In my fieldwork many conversations centred on questions raised by participant observation. It was through their explication in language that many observations became meaningful to me. At other times, I wondered if the knowledge expressed towards my assistant Suresh and me would be termed in the fishermen's everyday speech. Would they not try to make themselves understandable to a foreigner and an inlander 'from the east'[7], who both lack any ties to the fishing community and, in doing so, deviate from their usage of everyday language? In order to take this possible problem into account I attempted to generate conversation circles at the beach, trying to bring together people with contrasting or possibly supplementing opinions and experiences. Sometimes this meant disturbing men who were playing cards and often it resulted in my quickly becoming the onlooker of a conversation drifting away from my initial enquiry, myself unable to steer its course. But at other times it worked, and the moments when it did were some of the most rewarding of my work. Studying such conversations with detailed word-by-word transcriptions of the audiocassette recordings gave more reliable insights into the choice of words, phrases and arguments that had been articulated therein. These transcriptions also revealed the use of figurative language, which may or may not have been conscious to the speaker at the moment and which Suresh and I made the subject of further enquiries. In other instances I found that linguistic and non-linguistic knowledges supplemented each other – for example, when the fishermen were watching the surface of the sea and debated on their observations in order to reach a decision whether or not to embark on a fishing trip, or during a fishing trip when deciding on its continuation. Thus, language is an important element in the activity of fishing, not least because it is required in coordinating a team of thirty or so men in the large boats now in use.

Figurative Language

In retrospect, I tend to agree with Paul Ricoeur (1985: I–II) that a study can only consider what can be made conscious and that what can be made conscious can also be expressed in language. While I agree that some knowledge may per se be non-linguistic and non-sentential, much of it can be expressed in language nevertheless. As Black (1962), Ricoeur (1985) and Basso (1996) point out, much of cognitive thought may be revealed by figurative language, i.e. metaphors and other tropes,[8] whose study became a focal part of my fieldwork in Chamakkala. In Chapters 4 and 5, I shall deal more comprehensively with some different views on how figurative language may be interpreted. Here I shall mention only the 'interaction view' of metaphor proposed by Black, i.e. the idea that metaphors relate 'systems of associated commonplaces' with each other, resulting in the 'inter-animation' of two concepts (1962: 40). The theory presupposes that a metaphor always involves the interaction between a literal element in a sentence (the 'frame') and a metaphorical element (the 'focus'; ibid.: 28). Paying attention to metaphor may reveal such 'associated commonplaces' and bring to attention what may (or could) otherwise not be expressed in language. Keith Basso writes: 'It is in metaphor – perhaps more dramatically than in any other form of symbolic expression – that language and culture come together and display their fundamental inseparability' (1990:79).

Metaphors are meant to communicate, and as such they must be understandable to the members of a social group (Hesse 1980: 115). They may reveal concepts that may otherwise remain hidden. As Basso remarks, to elucidate the large bodies of knowledge hidden in these metaphors 'requires ethnography as much as it does linguistics' (1996: 69). Some examples are found in the conversations at the beginning of this chapter, namely the physiological character of the sea, its structure and apparent agency. On a more abstract level, the study of metaphors raises some profound conceptual issues. In particular, the distinction between literal and the figurative language which scholars like Black assume as given is often ambiguous and may change over time. As Valentine Daniel, writing on the speech of Tamil villagers, noticed: 'the line that divides the figurative from the literal is a thin and fragile one' (1984: 106). This is of importance for the interpretation of human-environment relations and knowledge change.

Memory

The conversation between Krishnankutti and Gangadharan at the beginning of this chapter draws attention to the important role of memory when considering 'local knowledge' and its transformation. As noted above, much of knowledge is revealed in the recollection of past events. Furthermore, to study its transformation (explicitly relevant in comparing the 'old times' with the 'time of progress' to which the two old men allude), one is relying on people's memories of the past.

At the moment of recalling the past, people themselves are living in the present. Thus, the process of memorising condenses the past and the present into the present. An example is the conversation of Krishnankutti and Gangadharan, who, while talking about the 'old times' kept switching the time between past tense and present tense. One may guess how much their memories of the fishing process were 'alive' and were virtually relived during the conversation. Indeed, their conversation appears like an emergent process in which memorising became a sort of performance. On this process, Barbara Myerhoff writes: 'At its most extreme form, memory may offer the opportunity not merely to recall the past but to relive it, in all its original freshness, unaltered by intervening change and reflection. All the accompanying sensations, emotions, and associations of the first occurrence are recovered and the past is recaptured' (1992: 238).

If, at its extreme, memory may retain its original freshness, it will usually be linked to the particular situation in which memories are retold. For example, recent events may be alluded to (and commented upon) and the composition of the audience may play a considerable role as well. As such, memory becomes a 'collaboration' of past and present, as Jonathan Boyarin writes: 'Memory is neither something preexistent and dormant in the past nor a projection from the present, but a potential for creative collaboration between present consciousness and the experience and expression of the past' (1994: 22). In this sense, the transformative and creative potential of memory is stressed, which may result in multiple, divergent accounts of specific historical events (Gold and Gujar 2002: 82–84). Talk about the past may reveal much about perceptions of the present. Yet memory is also embodied, as the above conversation between Krishnankutti and Gangadharan shows, who both refer to the bodily constitution of older fishermen, whose former strength is still imaginable. Similarly, in her work on identity among the Vezo, a fishing people of Madagascar, Rita Astuti (1995) notices that the identity of 'being Vezo' is inscribed in the body: a specific gait, scars resulting from fishing with lines and so on (ibid.: 40–41). The members of other communities living nearby do not share these features.

What is the range of the memorised past which I may hope to explore in this book? During my fieldwork the oldest fishermen in Chamakkala were in their mid-eighties. If so, they were born around 1915 and would surely remember events in the late 1920s. They remember the time prior to the Temple Entry Act of 1936, when Arayas and other members of low-status castes were allowed to enter large temples run by Brahmins. Examples of other 'markers of time', which seem to structure the chronological memories of many people, are a large storm and flood in 1937, the Second World War, a ban on the fishery of oil sardines (1943–1947), India's independence (1947), the formation of Kerala state (1956), the introduction of the *vaḷḷaṃ*-s (mid-1980s) and the different governments, political scandals, movies and their heroines and cricket champions.

Doing Fieldwork in Chamakkala

The fieldwork for this book was conducted between March 1999 and September 2000 and was supplemented by a revisit in late 2002. Initially, I stayed in Thiruvananthapuram, the state capital of Kerala, in order to improve my skills in the Malayalam language, taking daily lessons from V.R. Prabodhachandran Nayar, a retired professor of linguistics at Kerala University. In Thiruvananthapuram I met with fisheries experts and activists. During this time, I visited a number of coastal villages in Kerala, from the Tamil Nadu border in the south to Thalassery in the north. Eventually I decided to conduct my fieldwork in Chamakkala, a village in Thrissur district. Chamakkala seemed a particularly well-suited place for my work. The majority of fishermen in the village are Hindus, there has been a radical change in the fishing technology and there is a profound impact of work migration to the Arabian Gulf. Moreover, overfishing of the marine resources has proved to be a considerable problem locally. Most important, however, was the open-minded interest of V.K. Ravindran, a retired schoolteacher, and K.K. Subramoniam, a retired *tahsildar* (revenue subcollector), whom I had met through friends and who were to become the patron saints of my stay in the village. Some other people, including several fishermen, proved equally open-minded and helpful right from the beginning. I had first visited Chamakkala in May 1999, prior to the onset of the south-west monsoon, and returned a few times during the monsoon, since I found the phenomenon of the formation of *cākara*, a coastal mudbank appearing seasonally during the south-west monsoon, particularly intriguing. In October 1999, following a brief visit to Germany, I moved to a small house in Chentrappini which Ravindran had found for me. Previously, it had been the local headquarters of the CPI(M), the Communist Party of India (Marxist). It is situated in a part of the village that is jokingly called 'Moscow'. We had to remove a large pile of red flags from my 'Kremlin' before I could move in. Now my fieldwork was to begin.

While making first contacts with the fishermen and other locals I searched for an assistant to help me with translating and transcribing the language. I was capable of doing, and talking about, many everyday chores, but I kept getting lost in complex conversations throughout my stay. I was lucky that M.K. Sureshkumar became my co-worker. At forty-two years of age a married man, father of a son, and former businessman, Suresh came from the eastern part of the village. Belonging to the Izhava caste he knew none of the fishermen personally. However, thirty years ago he had gone to Chamakkala secondary school, situated near the beach. For him a number of memories of that time became alive during our work, and almost forgotten friends or their relatives became our interview partners. Suresh contributed much personal enthusiasm, motivation, skill and patience.

Our fieldwork was meant to consist largely of participant observation, participation in fishing trips (these were joined by myself without Suresh) and conversations with villagers, many of them informal chats but others systematic, tape-recorded interviews. The interviews were often meant to follow up on what Suresh and I saw. These conversations and interviews were not restricted to Chamakkala, but extended twenty or thirty kilometres up and down the coast. But our daily routine was in the village near the beach. It seemed to me that Suresh and I had soon become part of ordinary life. In retrospect, our fieldwork seems like a chain of different cycles of conversations and discussions, in which Suresh and I and many people of the nearby coastal region were soon enmeshed. For example, the three conversations with Krishnankutti that I described at the beginning of this chapter were part of a process of communication with him that lasted through the entire fieldwork. In this spirit, the conversations of the fieldwork of Suresh and myself extended much beyond prearranged gatherings for an interview. It included visits to Kumaran when he was mending the nets, watching the boats leaving the beach with Vasu, meeting the crew after return from fishing, sitting in front of Vijayan's house, and talking with Venu on an evening stroll to the beach, and so on.

Our fieldwork consisted of what Gudeman and Rivera (1990: 7) have called 'shorthand' conversations and 'longhand' conversations. The 'shorthand' conversations were the many conversations between Suresh and myself in which we reflected upon sights along the way and previous conversations. They took place along the way while walking or cycling through the village, when stopping for a little break in a tea-shop, when making endless transcriptions of interview tapes or while watching rituals. I tried to sharpen his attention towards the focus of my intended research. However, during the first months of our work I was often not sure what would actually relate to the 'local environmental knowledge'. We had to keep our eyes and ears open in all directions. Sometimes a specific ritual seemed interesting to me while another did not. I was often at pains to explain why this was so. Trying to explain became a permanent exercise to clarify my own thoughts. Between Suresh and myself the roles were clearly defined, but to outsiders we seemed like the 'research team': twins who were melting into one imaginary person. When I was seen in the village without Suresh they asked for him, if he was seen without me they asked for me. As such, the 'shorthand' conversations were 'inner dialogues' of this imaginary researcher-person.

What I shall call the 'longhand' conversations are those with our informants (or instructors, as I might call them as well). Through our enduring presence in the village a number of men became our primary informants. We would often return to them with new questions, for another interview, a brief chat or a cup of tea. In time the circle of informants increased. New questions led to additional people, and some of the men who had been reluctant to talk with us initially dared to talk with us later on, some only in the last weeks of fieldwork. Others, however, who had been enthusiastic about my work when I arrived, lost interest in it.

After six months in the village, prior to the onset of the south-west monsoon, I moved to the beach itself. I was hesitating to leave 'Moscow' where I felt comfortable with my neighbours. Moreover, there I could speak English with Ravindran and a few other men. But, with a local election coming up, the house was needed again for the activities of the Marxist party. When I had moved to the beach I was surprised by how much the move improved my work. First of all, it brought me much closer to the people at the beach in a personal way. Previously I had peeked into their world only in between 9 a.m. and about 8 p.m., except for the days when I joined fishing trips or watched activities on the beach in the early morning or attended a ritual at night. Living next door now, I was subject to children's curiosity, but also to my neighbours' friendship and generosity. A neighbour's wife, N.H. Thulasi, prepared my food and washed my clothes. Sometimes I would eat with her family. A kind and warm-hearted woman, Thulasi persuaded some of the women from the beach area to talk with Suresh and myself as well. Living near the beach was also revealing for watching people of the different communities (Muslim, Hindu Araya, Izhava and Vettuva) interact with each other. Moreover, since the monsoon is the most important season for fishing, living at the beach meant to be more up to date with current affairs. And, eventually, I became myself in an unanticipated way. As noted above, Suresh and I had soon become the 'research team' and were considered as an imaginary single person. This was often funny, but sometimes uncomfortable for both of us. By living at the beach I became the permanent resident and Suresh the daily visitor. Thus, the 'researcher-person' became a more differentiated being.

Nevertheless, sometimes my nerves got the better of me: suddenly I had a feeling of claustrophobia and the desire to be just a plain normal person on the road, not being constantly watched by the people around. Such feelings could be remedied in part by a visit to the district capital Thrissur. What made my journeys to Thrissur particularly worthwhile were my visits to L.S. Rajagopalan, a Tamil Brahmin scholar living in an old wooden house in the Punkunnam Agraharam (a Brahmin neighbourhood). Not only did he help me with finding contacts and making copious translations, he also cheered me up with his good humour.

After my move to the beach, the work of Suresh and me proceeded more or less in the same way as before. He would come in the morning to my house. Then we decided whether to go on a tour round the beach, visit some men, or have an interview, or Suresh would do transcription work in the house while I continued with the household census or the boat survey.

The cooperation between Suresh and me was not always smooth. At several points during fieldwork we had serious conflicts and were once on the verge of discontinuing our joint work altogether. When these tensions seemed to surface all at once, in March 2000, Suresh and I took a break for a few days and then sat together to discuss our situation. We quickly realised that it was impossible for either of us to stop our cooperation at this point. We both knew that our work had gone smoothly most of the time and that, altogether, we got along quite well personally. The remedy for our tensions was to stick closer to fixed working hours (so that Suresh could return to his family in the evening in good time) and that I gave more room to Suresh for being, and appearing as, the co-worker. The ensuing months of fieldwork went along smoothly. There was no other clash of this magnitude and I believe that we had eventually found our pace. Or was it the knowledge that the end of our intense work was approaching that made us both feel lighter?

In late September 2000, when my year in Chamakkala was over, I was picked up by friends with their car to take me to the Thiruvananthapuram train. I was stunned. A few hugs and handshakes among the silent crowd that had gathered at my house at the beach was all I was capable of. I was happy to leave, but I was unhappy to leave them, the people who were so open-minded, helpful and humorous to a stranger like myself.

Notes

1. Throughout this book, I shall write personal names in a 'western style': given name first, house (i.e. lineage) name second. In this way individuals can be identified more readily. This differs from the official Indian style of writing names, where the given name is listed after the abbreviated names of the lineage and of the father.
2. Later I realised that most men emphasise the mountain water, rather than the mountains, as the partner in the intercourse. I could not find any similar Indian myth in the literature. However, conceptually it somewhat resembles the relationship between the Earth goddess Prithvi and the sky god Indra. According to Caldwell, in the Rig Veda, Prithvi is described as 'the ready, heated female body, seeking the soothing and fertilising seminal rains of the sky-god Indra' (Caldwell 1999: 104; see also Dossi 1998: 20–29).
3. This term is one of a confusing variety of terms recently employed by anthropologists and others to describe and refer to the subject(s) of their study: 'indigenous knowledge', 'indigenous technical knowledge', 'ethnoecology', 'local knowledge', 'localised knowledge', 'art de la localité', 'folk knowledge', 'traditional knowledge', 'traditional environmental (or ecological) knowledge', 'people's science' and so on (Ellen and Harris 2000: 2). Ellen and Harris note that, while these terms are often used interchangeably

(assuming that they will be understood in a 'common-sensical' fashion), their meaning is not identical (2000: 2). Moreover, what is 'indigenous', 'traditional', 'local', 'folk', 'knowledge' and 'science' means different things to different people in different contexts.

4. In the Western calendar, the fifteenth day of the (Malayalam) *iṭavaṃ* month is approximately 1 June, the tenth day of the *karkkaṭakaṃ* month is in late July. The months of the Malayalam calendar are listed in the glossary at the end of this book.
5. Inasmuch as knowledge is contained in stories or recollections of past events (which it often is), an understanding of this knowledge must be closely related to an understanding of memory.
6. According to Bloch 'mental models ... integrate visual imagery, other sensory cognition, the cognitive aspects of learned practices, evaluations, memories of sensations, and memories of typical examples' (1992: 130).
7. Coming from the eastern part of the village, Suresh was often termed a *kiḻakkāran*, literally a 'man of the east'.
8. In the Introduction of his book *Metahistory*, Hayden White (1973) provides an intriguing elaboration of the four classical tropes metaphor, metonymy, synecdoche and irony.

2 West and East

Figure 2.1: Mending nets on the beach

If any place is central in the everyday life of fishermen in Chamakkala, it is on the beach, under the westernmost palm trees. Sheltered from the burning sun, both the sea and the paved road that links the beach with the inland can be seen from here. This is the place where the men convene to discuss conditions in the sea, whether or not to embark on a fishing trip, to debate joint business or mend their nets and to play cards or have a nap (Figure 2.1). This is also where affairs concerning the beach (*kaṭavu*) are negotiated. This is the place where government officials come to instruct fishermen about its programmes, and this is where the local branch of the *dīvara sabha*, the Hindu fisherfolk's caste organisation, holds its meetings. Most of the time, somebody is sitting at or near this place. But even in the absence of people its central significance can be guessed from the large number of cigarette butts and used matches, remains of lottery tickets and torn playing cards lying in the sand. I never saw a woman sitting at or near the 'convention place'. The beach is a male space in general and this is the most male-dominated place of all.

Figure 2.2: The bhaṇḍāraṃ *box on the beach in Chamakkala. The upper inscription reads* dēvisahāyaṃ *('help/protection of the goddess'), alluding to Palappetti Bhagavathi, the beach's* stala devata *(lit. 'place deity').*

The boats of Chamakkala are clustered here. Nearby, nets are spread out for drying or repair. A *bhaṇḍāraṃ* box is also in sight (Figure 2.2). This is a small column made of concrete, in which money can be dropped as a donation to the temple at Palappetti, about three kilometres to the north. The temple's deity is Bhagavathi, the benevolent form of the goddess in the Hindu pantheon. Palappetti Bhagavathi is the *stala devata* (lit. 'place deity', derived from *stalaṃ*, place, and *devata*, deity) of the beach and the *bhaṇḍāraṃ* box is a visible sign for her residence. The money collected therein is donated for the temple's annual festival. Many fishermen drop in some coins before entering their boats.

Approaching the Beach

Coming from inland, one reaches the beach of Chamakkala after leaving National Highway 17 (Mumbai – Kochi) at Chentrappini Centre in Kerala's Thrissur District, turning westwards. Chentrappini is an extended, non-nucleated village stretching over most of a *panchayat*.[1] It is situated in between the Arabian Sea in the west and Canoli Canal in the east. Chamakkala is Chentrappini's coastal ward. Spreading along the highway in a bustling semi-urban sprawl, Chentrappini Centre looks much like the other villages in the coastal region. People from the area come here to buy goods, go to the post office, meet a doctor or catch a bus at the bus stop. Two- and three-storey concrete buildings are home to new and old businesses. They have largely replaced the tiled Malabar-style houses with their wooden walls and wide verandas.

Taking the road to the west, the bustle soon gives way to lush, green garden land with houses sprinkled in between. Coconut palms are abundant; bananas, vegetables, papaya, jackfruit, tapioca and spices are grown as well. Here and there Malabar-style houses are crouching under the coconut palms. Arabian- and North American-inspired villas (all inhabited by the families of Gulf migrants) are seen on larger compounds. About one kilometre west of the highway, there is a small cluster of brick-built houses at a junction. A ration shop[2] sells basic food items such as rice, lentils, salt, sugar, tea and spices at subsidised prices. It also sells kerosene, used not only for cooking in private homes but also as fuel for the outboard motors of the fishermen's boats. There is a butcher, a mill, a fabric store and a fruit and vegetable stall. The next junction, perhaps two kilometres west of the highway, is called 'Four Cross'. This is where Chamakkala begins (see Map 2). There is a small mosque. Next to it a *madrasa* (Muslim religious school) is under construction. There is also a tea-shop, a provision store, a shop for household items, a fruit stall, a tailor, a hairdresser and a small office building, with a doctor in the basement and the office of a local fisheries cooperative upstairs. A few hundred metres further west, there is another junction, with Chamakkala secondary school, Chamakkala mosque and its attached *madrasa*. There is also a small post office, a tea-shop, a provision store, a rice mill and a store for household items. A Hindu temple, belonging to an Araya lineage, is hidden behind these stores. The road passes the mosque and the Muslim cemetery. Further towards the west there are several concrete villas, all owned by Muslims.[3] At 'Three Cross' junction, West Tippu Highway[4] is reached. This road links Chamakkala to the other beaches further north and south. At 'Three Cross' there is a stop for the buses connecting the coastal villages, the office of a cooperative bank, two telephone booths, an English pharmacy (selling cosmopolitan medicine), a hairdresser, two tailors (one with an attached duty-paid shop, where imports from the Arabian Gulf can be bought), a provision store, a basic restaurant and two cycle repair shops. Nearby is the large plot of the Al-Mullah child orphanage, a small mosque and another *madrasa*.

Turning north on West Tippu Highway, one would reach the Primary Health Centre (PHC; since 1988 staffed with a permanent doctor and four nurses). Turning south on the highway, about half a kilometre south of 'Three Cross', the beach road branches off towards the west. Along this road and beyond there are small compounds with either small tiled, brick-built houses or simple huts, thatched with palm leaves. The area is much more congested than further inland. Compounds are separated by fences made of palm leaves rather than the concrete walls seen further east. Sparse vegetation grows on the sandy soil. The road passes a small temple compound and, across the road, the only two-storey concrete building in the beach area. There is another small provision store. After two hundred metres the road ends at the beach near a *bhanḍāraṃ* box. Nearby there is a shelter for fishing equipment and a thatched storage house for dried and salted fish. In between the coconut palms nets are spread out for drying and for repair. Some fishermen mend the nets, others chat, play cards, watch the sea or sleep. A few steps further west the boats are seen lined up on the beach.

The part of the shore belonging to a village is known as a *kaṭavu* ('jetty'). It is the spatial unit of local fisheries. Fishermen, middlemen (*tarakan*-s) and their boats belong to specific *kaṭavu*-s. The Chamakkala *kaṭavu* is limited by two small streams, which enter the sea in the north and in the south in two inlets (sing. *aṟappa*). These are approximately one kilometre apart. The next *kaṭavu*-s to the north and south are called Kazhimbram and Companykatavu. They belong to the nearby villages Edamuttam and Kaipamangalam.

Appendix B details demographic data on the population living in the beach area, including the results of my census of 326 households.

Cardinal Points

Once the fluency of my Malayalam improved, I was able to ask people in Chamakkala about how to reach specific places – and to understand their answers. I noticed that they rarely use the terms left (*iṭattụ*) and right (*valattụ*) in describing the way. Instead, they refer to the cardinal points east (*kiḻakkụ*), west (*paṭiññāṟụ*), north (*vadakkan*) and south (*ṭekkụ*). For instance, when one asks how to reach the house of, say, Chandran (as in the sketch in Figure 2.3), the answer would not be to go straight, turn left at A, then go to B, turn right and go straight to Chandran's house. Rather, one would be told to go north to A, turn west, go to B, turn to the north again and thus reach Chandran's house. Spatial orientation makes use of an absolute frame of reference instead of a relative frame of reference intrinsic to the moving person (Levinson 2003).

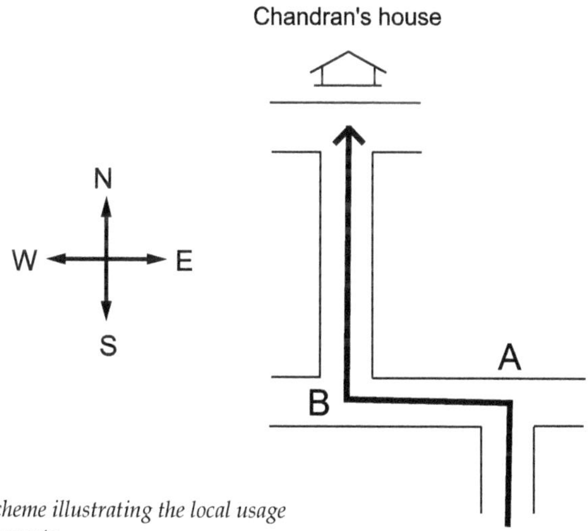

Figure 2.3: Scheme illustrating the local usage of deictic statements

As soon as I paid attention to the cardinal points I realised how much they were used in a variety of everyday situations. A few examples illustrate this point. When I met a few fishermen on the beach for a friendly chat, one of them offered me a seat on the western side of his mat (woven of areca leaves, measuring perhaps one by two metres). Another day, when two boats were brought from near the sea further inland, it was literally said to 'bring the boats to the east (*kiḻakkụ*)'. When one boat did not fit in between two coconut palms, it had to be taken a little bit further 'to the west' in order to find a better way to the east. This is an example of the use of the cardinal points in describing movements. It is relevant even on much smaller scales. When a net was once dyed in a small tank (measuring perhaps one metre in diameter), dye was poured on it from a can. It was stressed that the dye had to be poured 'from east to west'.

Another instance revealed to me that the referential use of the cardinal directions applies within houses as well. It also hints at the early age at which these meanings are learned. Binsi, the daughter of my neighbour Kuttan and his wife Ambika, was just over two years old when I lived on the beach. On a rainy day during the south-west monsoon, Kuttan, Ambika, Binsi and I were sitting in their palm-thatched hut in the evening. Their *vilakkụ*, a small oil lamp made of brass, was standing near the (eastern) entrance of the hut. It had just been used by Ambika for the *vilakkụ kānakkuka* rite, in which it is carried around in the house to secure the blessing of Laxmi, the Hindu deity of wealth and prosperity (see Moore 1990: 177). Kuttan was afraid that the rain would extinguish the lamp and thus he asked Binsi to move the lamp carefully a bit further 'to the west' (i.e. into the hut) – which is what Binsi did.

It was remarkable to me that villagers (including Suresh) could often tell the cardinal points when I happened to ask them and were able to use them for orientation. Though familiar with the use of the cardinal points as an absolute frame of reference in circumstances such as these, the extent to which fishermen in Chamakkala and other coastal villages in Thrissur District allude to them was surprising to Suresh and other Malayalis from inland who visited me or whom I told about this.

The cardinal points east and west are meaningful in a way that is altogether unfamiliar to Malayalis from inland. Chamakkala fishermen divide the coastal area into the west and the east and a small space in between. The west (*paṭiññāṟụ*) is the Arabian Sea (*arabi kaṭal*), the fishermen's environment for work. The east (*kiḻakkụ*) is the inland. The fishermen's area of residence is in between. Less emphasis is put on the meanings associated with the other cardinal points, north and south.[5] East and west are further differentiated: the nearby regions in these directions are referred to as *kiḻakkuppuṟaṃ* ('outside in the east') and *paṭiññāṟuppuṟaṃ* ('outside in the west'). The suffix *puṟaṃ* denotes 'outside'. While *paṭiññāṟụ* and *kiḻakkụ* denote directions, *paṭiññāṟuppuṟaṃ* and *kiḻakkuppuṟaṃ* denote spaces situated in these directions. The fishermen's residence on the beach is thus constituted as an 'inside', but it is not explicitly designated as such.[6] To associate spaces with the cardinal points was unusual for my co-worker Suresh, but is commonly done by fishermen, at least along the entire coastal strip between Chavakkadu and Kodungallur. For the fishermen, Suresh, who lives east of the National Highway in Chentrappini, is a *kiḻakkāran*, a 'man of the east'. For them there are the 'men of the east' (*kiḻakkāranmar*) and the 'men of the west' (*paṭiññāṟukkāranmar*). And there is the 'work of the east' (*kiḻakkan pani* or *kiḻakkan jōli*) and the 'work of the west' (*paṭiññāttụ pani*): fishing. To some extent, both *kiḻakkuppuṟaṃ* and *paṭiññāṟu(ppuṟaṃ)* depend on the places of references and constitute relational categories. In many conversations I have attempted to localise the border of the *kiḻakkuppuṟaṃ*. But I failed. It cannot be identified with specific places. On a larger scale, *kiḻakkuppuṟaṃ* and *paṭiññāṟu(ppuṟaṃ)* have an essentialist meaning. The west (*paṭiññāṟu*) is always the sea.[7] Chamakkala mosque and Four Cross junction do usually and Chentrappini Centre does certainly belong to the *kiḻakkupuṟaṃ*. Its essentialist character became evident to me when, while being near the National Highway in Chentrappini with a fisherman, he told me: 'Now this is the *kiḻakkuppuṟaṃ*!' In this essentialist sense, *kiḻakkuppuraṃ* and *paṭiññāṟu* constitute an (unnamed) 'inside' space (the fisherfolk's area of residence) and two (named) outside spaces.

A Community of Conversationalists

Figure 2.4: Young fishermen of the Ponnonam unit.

On a typical day of our fieldwork, Suresh and I went to the fishermen's 'convention place' first in order to catch up with the news. How many men were sitting here depended on the day's fishing activity. On a busy day, when middle-aged and younger men were out fishing, some elders would sit here, play cards, discuss weather conditions and recent catches and wait for the news about the day's catches and the boats to return.

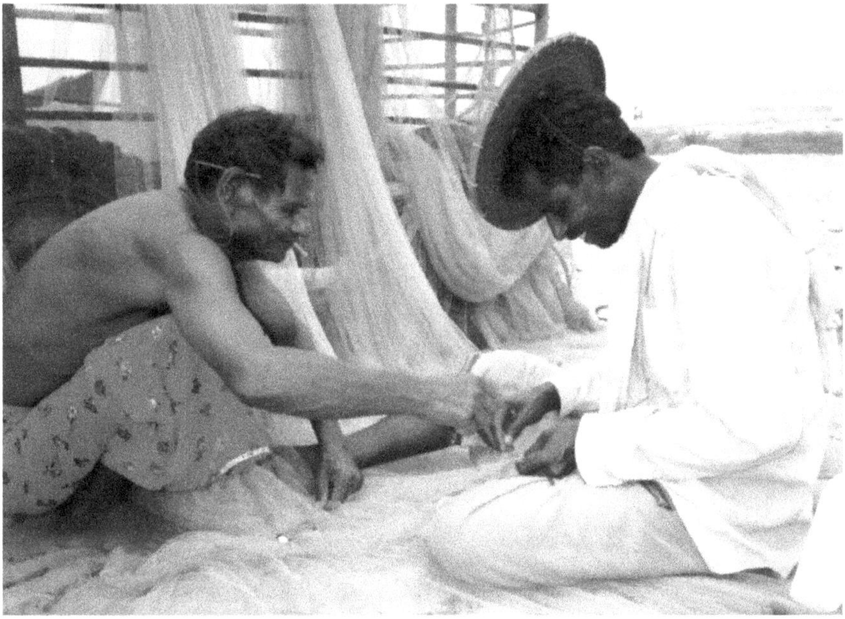

Figure 2.5: Vasu Eranpurakkal (left) and Balan Thaykattu (right) mending nets.

On idle days, when the boats had not left for fishing, a larger and more diverse group of people would be present (except for days of intense rainfall, when people dispersed to seek shelter under the veranda of the nearby provision store or any of the sheds). Besides middle-aged and older men who are often present here, there are young fishermen in the village as well, but on an idle day they rarely go to the beach (Figure 2.4). Instead, they prefer to meet further east, where the beach road meets West Tippu Sultan Highway. They may be trying to find work as a day labourer in the east, perhaps in agriculture or as a head-load worker. Or they may plan to go to a cinema in Chentrappini Centre. In doing so, these young men make a distinction between the beach as a sphere of work and the eastern inland (the *kiḻakkuppuṟaṃ*) as an ambivalent place of work and leisure activity. For older men, the spheres of work and leisure overlap. Not only do they work (mending nets, doing repair work) and play cards in the same space (Figure 2.5), but by sitting at the beach they also learn about changing conditions of the sea and the weather conditions. They draw inferences from watching other boats perform in the sea. These observations contribute positively to future success in fishing. It is the usefulness of being 'in touch' with the environment, a generally esteemed habit of hunter-gatherers (Bird-David 1992: 30; Ingold 2000: 47, 55).

Of the two men who have been dominating Chamakkala fisheries economically for many years as *tarakan*-s (middlemen), i.e. moneylenders-cum-auctioneers of fish, one often joins the group of card players and spends much time on the beach while the other is rarely seen there.

Most fishermen above the age of fifty or so years have worked in local fishing throughout their lives. Many younger fishermen have aspired to leave local fishing in search of employment in Madras and the Arabian Gulf. Returning to fishing after a stint in the Gulf is considered a failed attempt to achieve upward social mobility. In general, successful migrants to the Gulf do not return to work as fishermen. The few who do so typically become the owners of a boat and gear. There are some men in neighbouring villages who have experimented with new boat designs and fishing techniques after their return from the Gulf.

Occasionally, Gulf migrants from Chamakkala are seen on the beach on a visit home. Some of them even work for a few days on the boats of their former colleagues, symbolically reinforcing former ties. Others refrain from doing so and visit the beach only briefly, demonstrating upward mobility by wearing white *muṇṭu*-s (loin cloths) and white ironed shirts, and often gold imitation Rolex watches.

On weekends and afternoons, (male) students and unmarried younger fishermen transform two stretches of the beach into cricket grounds. As elsewhere in India, cricket is the most popular sport in the village. But, for most of the younger fishermen, Chentrappini Centre with its movie cinemas and soft drink shops is the more exciting destination. Following days of abundant catch, women may be seen laying out fishes in the sun for drying.

Not many Muslim men come to the beach, even though they are the largest community in the coastal ward. Most of those who do are fish vendors, but there are a few Muslim fishermen as well.

Hindu Arayas and Muslims in Chamakkala

Almost all of the Hindu fishermen in the village belong to the Araya caste,[8] one of three large Hindu fishing castes in Kerala. Marine fishing is the *kula toḻil* (lit. 'lineage work', traditional caste occupation) of Araya men. Kerala's other large fishing castes are the sea-fishing Mukkuvar and the backwater-fishing Valan. According to K.P. Padmanabha Menon (1993: 2ff.), the *Jāti Nirnnaya*, a Sanskrit poem explicating the caste system in Kerala, ranked the fishing castes as 'degraded castes' (*patita jāti*) below Brahmins, Shudra (including Nayars), intermediate castes (including temple servants) and artisans, but above 'polluted castes' (*nīca jāti*) like Parayans, Pulayars, Vettuva and tribals. The *Report on the Census of the Cochin State for 1901* ranks the fishing castes along with Izhavas, Kaniyans, Paanans and Veelans as castes whose members pollute Brahmins, Nayars and members of other 'pure castes' if approaching these within thirty-six feet (quoted by Padmanabha Menon 1993: 24).[9] Only castes of agricultural labourers (Parayans, Pulayars and Vettuva) and tribals were considered as being more polluting (at distances of 48 and 64 feet from Brahmins, respectively; ibid.: 24-26). It is difficult to ascertain how strictly these limits were ever followed. The anthropologist L. K. Ananthakrishna Iyer (himself a Brahmin) notes that Arayas rarely travelled inland. Not being allowed to travel along public roads they were allegedly obliged to keep to the sea coast (1909: 261, 265).[10] Thus, rules of distance pollution seem to have contributed to a spatial segregation of different castes.

The Araya self-image differs significantly from these attributions of low status. On my enquiry, many old men derived the caste name 'Araya' from the Sanskrit word *rajah* (king). Those men who did not claim direct descent from the Rajah of Kochi at least surmised that the Arayas must have been boatmen in direct service to the Rajah, and thus were inevitably of high status. In an elaborate exposition, an Araya teacher derived the names of several South Indian high-status castes such as Nayars and Tamil Brahmins (whose common lineage name is Iyer) as corruptions of the word 'Araya'. As such, he claimed a high status for the Arayas and regarded it as a fact that had somehow fallen into oblivion. Besides idiosyncratic interpretations such as these, a myth about the alleged Brahmin origin of the Araya caste is widely known among older men. Kumaran Eranpurakkal, a fisherman in his mid-seventies, explains:

Kumaran Eranpurakkal: According to a legend (*eitihyaṃ*) we were Brahmins. It was said like that. But once a banquet was held [of all the Brahmins of Kerala] and tender mango pickles were served. When somebody accidentally squeezed a mango, a seed came darting out. An Araya man said spontaneously: 'Ha! That looks just like a small shark dashing away!' Everybody understood that this man had been fishing in the sea. It was clear that the Arayas had to be separated (inf. *tirikkuka*) from the other Brahmins.

Götz Hoeppe: Why was it necessary to separate them?

Kumaran: In the olden days it was done like that. It was done according to the rules (pl. *niyamaṅṅaḷ*) of that time. But if we visit a Nambudiri [Brahmin] scholar now and tell him that we are from this [Araya] community (*sammudāyaṃ*), then he will get a chair for us to sit down on. This is the [respectful] way in which they behave towards us.

Several people told me that there is factual evidence for this Brahmin origin. At Araya funerals, the priest wears the sacred thread (*punnūlụ*), which is usually worn only by members of Brahmin and Viswakarma (artisan) castes. Legend has it that Araya men used to wear the sacred thread[11] all the time. But while fishing in the sea either it was found inconvenient (and therefore taken off) or the thread's material was used up in mending the net.

One may note that to make a claim to a former high social status implicitly means to concede the caste's sunken status in the present. Given the rules of distance pollution mentioned above and the alleged prohibition of Arayas to walk on public roads, these claims to high caste status seem like a 'hidden transcript' of the subaltern, a denial of the dominant public discourse (Scott 1990). That Arayas did feel much denigrated was most clear when I talked with older men about the religious practices in their childhood. The prohibition of Hindu *avarna* caste members to enter *mahakṣētraṃ*-s (large temples run by Brahmin priests) was lifted only in 1936 when the Temple Entry Act was passed (Menon 1994: 106–117). It is painfully remembered by old Arayas. Except for the annual festival, these temples had been closed to them. Offerings to Palappetti Bhagavathi, the *stala devata* (local deity) of the beach and the fisherfolk's *para devata* (supreme deity), had to be passed on to high-status caste Hindus or the priests in order to be brought to the deity.[12]

Social mobility and occupational differentiation of Arayas has increased since India's independence in 1947. In 1975, all Hindu fishing castes in Kerala were joined in a common statewide caste organisation, the *dīvara sabha* (lit. 'assembly of the clever ones').[13] It aims at representing the fishermen politically and coordinates claims and attempts to raise their standard of living and social status.[14] There is a local *sanghaṃ* (assembly) of the *dīvara sabha* in Chamakkala.

There has been a tendency of segregation between fishing and non-fishing Arayas. This is recognisable in the marriage register,[15] which reveals that daughters of fishermen are usually married to fishermen, while daughters of non-fishermen (who work in the Gulf or as schoolteachers, clerks or moneylenders) are married to men who are themselves migrants to the Gulf or who do work that is considered to be of a status superior to fishing (e.g. work in an office or a shop). The difference between lineages with fishing and non-fishing Arayas has become noticeable in some lineage temples. There is one lineage temple in Chamakkala of a house with many non-fishermen, many formerly or still employed in the Gulf. In this temple, non-vegetarian offerings to the main deity, Bhadrakali, were abandoned in favour of vegetarian offerings several years ago. At about the same time, it was decided to conduct *pūja*-s in Sanskrit. In contrast, there is a branch of this temple near the beach, attended by fishing house members only, where non-vegetarian (blood) offerings are made to Bhadrakali and the *pūja* is conducted largely in Malayalam, the local vernacular. The same tendency can be observed for other Araya temples in the area. The use of Sanskrit and of vegetarian (i.e. non-violent) offerings alludes to a (high-status) Brahminical style of worship and underlines the claim to upward social mobility of the worshippers (see Osella and Osella 2000: Chapter 5).

The descendants of a former Araya headman (*kadakkoṭi*[16]) are among the Arayas who emphasise their successful disengagement from fishing. One of them told me in English: 'There are two kinds of Arayas: the advanced Arayas and the backward Arayas. The advanced Arayas! We are doctors, teachers and IAS officers.[17] But the fishermen, they are illiterate and backward'.

Gulf migration has made possible considerable rises in a family's economic and educational standing within one or two generations. Reservations granted through the OBC ('other backward castes') status were introduced in the 1970s and have made it somewhat easier for Araya men and women to achieve higher education and secure government employment. Yet most consider employment in the Arabian Gulf, often as unskilled labourers, a more realistic means to achieve upward social mobility. The best chance, available to few men only, is employment in a fibreglass factory in Ajuman (United Arab Emirates), which is owned and run by an Araya man from Chamakkala. Most employees are male members of his lineage.

While for the Arayas fishing as the 'lineage occupation' (*kula toḻil*) is the most obvious factor constituting their caste identity, Muslims do not define themselves through their involvement in the fisheries. Even before migration to the Arabian Gulf became significant in the 1970s, Muslims were apparently employed in a wider range of occupations than Arayas, most being of low status (including agricultural labour). The local Muslim[18] population appears to have three distinct origins: Arabian traders who were in contact with the Malabar coast since shortly after the foundation of Islam in the seventh century (Miller 1992: 39–46), people who deliberately converted to Islam (ibid.: 46ff.), many stemming from low-status Hindu castes, particularly fishermen (ibid.: 55–56), and people forcibly converted to Islam during the Mysorean invasion in the late eighteenth century (see Dale 1980: 79–91; Miller 1992: 85-94; Gabriel 1996: 163–237).[19]

Despite their various possible origins, the Muslims of Chamakkala belong to a single community of approximately five hundred households. The large mosque, its two branches at the Three Cross and Four Cross junctions and the attached *madrasa*-s are part thereof. Formally, these Muslims are Sunnis who follow the Shafi'ii school.[20] The local practice of Islam includes the worship of Muslim saints, the most revered being Mohiuddin Sheikh.[21] Recitals of *mavlūdụ* texts (praises of the prophet and of Muslim saints) and vow offerings to Muslim saints (*nērca*-s) are, or have been, common. Occasionally, *ṟatēbụ*-s (rituals in which Muslim saints are evoked in possession) were conducted. In recent years these rituals have been subjected to increasing criticism from Mujahid, a reformist movement emphasising a strict following of the Quran (Ahmad Kutty 1995). Its influence is growing, perhaps due to migration to the Arabian Gulf. There is a Mujahid mosque in Kaipamangalam, the village south of Chamakkala.

The increasing contact of local Muslims with Arabian Islam manifests itself visibly in the architectural designs of local mosques. Until early in the twentieth century, mosques in Malabar resembled Hindu temples and Christian churches. These were wooden buildings with tiled roofs. The new mosque of Chamakkala deviates from this style. Made of concrete and supplemented by a minaret tower, topped by onion-shaped domes, it alludes to an Arabian architectural style. It is said that the construction of the new mosque has been possible thanks to the donation of a sheikh from Dubai. The mosques in Chentrappini Centre and at Four Cross have been built in a similar style. Both were constructed recently.

Older Muslims and Arayas in Chamakkala commonly stress the good relationship between their communities. In arguing this point, emphasis is put on the shared involvement in fishing and fish vending in the past. It is remembered by Muslims and Arayas alike that Muslims were given fish by Araya fishermen when the village was struck by poverty.

There are traces of a former local religiosity. In particular, a number of Muslim and Hindu men regard the Muslim saint Mohiuddin Sheikh to be identical with the Hindu deity Subramoniam. This identity is suggested to them by the similarity in ritual, especially the conduct of possession rituals. Both the *veṭṭuṃ kuṭṭuṃ ratēbụ* (done in praise of Mohiuddin Sheikh) and the *tuḷḷal* (possession, lit. 'dance'; done in praise of Subramoniam) involve bodily mutilations: the former the cutting and scratching of skin, the latter the piercing of a *śōlaṃ* (spear) through the tongue of the possessor. Unlike Mohiuddin Sheikh, Subramoniam is not considered to maintain a special relationship with the sea. However, a number of fishermen name Subramoniam as their *iśṭa devata* ('chosen deity'). The pilgrimage to the most famous Subramoniam temple in Palani (Tamil Nadu) has an ever-growing attendance.

The perceived identity of a Muslim saint and a Hindu deity parallels the notion, widespread in coastal central Kerala, that Allah resembles Lord Shiva in being the supreme deity and that, prior to becoming a centre of Muslim faith, the Kaaba in Mecca (Saudi Arabia) had been a *bāna* ('arrow') *lingaṃ* and, therefore, its surrounding mosque a Shiva temple. Arguments supporting this identity include the similarity in the dress worn by *haji*-s (pilgrims to Mecca) and that of Brahmins visiting major Hindu temples, as well as the resemblance of the Kaaba to a Shiva *lingaṃ*. Older Muslims and Hindus share this notion, but a growing number of younger Muslims began opposing it vehemently.

In recent years, Gulf migration has led to a considerable disengagement of Muslims from fishing-related work in favour of better-paid and higher-status work. Among many Arayas this has created a feeling of being disadvantaged or even marginalised by Muslims who were once marginalised as coastal dwellers themselves. With the opening of the secondary school in the early 1970s and the post office near the main mosque in the 1980s, 'Chamakkala' became the official name for the seaside ward. The name 'Chamakkala' (lit. 'millet stub' after, *cāma*, millet, and *kala*, stub) is derived from the lineage name of a locally influential Muslim family of possible Brahmin origin. Until then, the Arayas used to call their beach (*kaṭavu*) and the adjacent land 'Karimbram'. Nobody uses this name any more. Some Arayas in Chamakkala also feel marginalised politically by their Muslim neighbours. As there is a solid Muslim majority in the ward, in recent years the ward representatives at the *panchayat* council have always been members of the Muslim League.[22]

Araya men consider work in a *cappa* and, less so, fish vending, as activities that are inferior to fishing. Both occupations are associated with Muslims: 'Hindus would not do such a work which involves these strong smells. I can only do my business thanks to my Muslim workers', says an Araya *cappa* owner and *tarakan*. Indeed, all men at present or previously involved in salting fish are Muslims. Some Araya women also partake in this work. Several people from inland told me that previously Muslims and Arayas from the beach area were considered as being alike in that they smell of fish. A derogatory saying addresses people who are accused of being unwilling to do serious work: 'Can you not go and sell oil sardines?' (*ninakkụ cāla vilkkān pōyikkuṭe*?). With this insult the common occupation of Muslims and Arayas is depreciated.

There can be little doubt that local Hindu–Muslim relations have been deteriorating lately. In recent years there have been clashes in Chentrappini and Chamakkala between younger Muslims and Hindus on the anniversary of the destruction of the Babari mosque in Ayodhya (6 December 1992; see van der Veer 1994). After the 1992 riots, several Araya men who had succeeded in finding work in the Gulf were laid off by their Muslim employers. While several unskilled Muslims have found employment in the Gulf since 1992, it has been very difficult for similarly unskilled Hindus to achieve the same.

Within the last few years, local groups of the Hindu nationalist party BJP (Bharatiya Janata Party) and the RSS (Rashtriya Swayamsevak Sangh), a youth organisation with strong emphasis on martial practice, have drawn an increasing number of followers among Hindus in Chamakkala. One contributing factor seems to be the economic success and apparent unity of local Muslims. Feeling disempowered *locally*, several Hindus seek to align themselves with Hindu fundamentalist movements, which are more powerful and which articulate Hindu *nationalist* positions, as RSS and BJP do.

Gulf Migration and Muslim Disengagement from Fishing

Work migration to the Arabian Gulf is the most important single factor that has brought about the prospect of upward social mobility in Chamakkala. The inflow of money has affected all spheres of living, including fishing. Within a few years migration has contributed to a considerable difference in wealth between Muslims and Arayas. Among Muslims, the number of Gulf migrants per household is three times that in Araya households (Appendix B, Table B.2).

Migration from Chamakkala more or less matches the Kerala profile, with regard to both its historical development as well as in its structural properties (see P. A. Kurien 1994; Isaac 1997; Osella and Osella 2000: 77–79; Parayil 2000). Historically, outmigration of Chamakkala men dates back to the 1930s.[23] At that time, some of them found work in Sri Lanka and Assam, but most had returned by the 1950s (see Isaac 1997: 270). In the 1960s and 1970s, Madras and Bombay were popular destinations of migration. While trade links between Malabar and the Arabian Gulf had existed for many centuries, massive migration to the Gulf began in the early 1970s (ibid.). The number of migrants from India to this region peaked in the 1980s, but declined sharply thereafter, especially after the Gulf war of 1991. My household census shows that in early 2000 Gulf migrants formed 95 per cent (188 of 200) of the Chamakkala males working outside Kerala (Appendix B). There was not a single female migrant from the village. Structurally, migration of Chamakkala men to Arabian Gulf states is determined by the migrants' qualifications, the form of their work contracts and by the local community to which they belong. Having often left secondary school without obtaining a Secondary School Leaving Certificate (SSLC) and lacking professional training, 82 per cent of the Gulf migrants from Chamakkala (155 of 188) are engaged in unskilled manual labour, such as construction work, work as cleaning staff, packing and unpacking boxes on a market or, rarely, fishing. Skilled or professionally employed Gulf migrants from Chamakkala work as accountants, typists, in technical drawing or even as engineers.

The survey suggests that the onset of Gulf migration has gone along with the disengagement of Muslims from fishing-related occupations. As mentioned above, there are today fish vendors in 20 per cent of Muslim households in the survey area (31 of 155). On asking in all Muslim households about the relation to fishing and fish vending, I was often told that there is no such relation. However, when I enquired about the father's (or, more generally, the previous householder's) occupation, I learned that in 103 households (66 per cent) he had been a fisherman or fish vendor.

In 52 households (34 per cent) this was negated. Assuming that this fraction was constant during the last generation (note, however, that the number of households must have been smaller because of the partition of property with time), it seems that a majority of Muslim households had been involved in fish trading or fishing previously. Therefore, local Muslims must have considerably disengaged themselves from fishing-related occupations within one generation.[24]

In the Gulf, remuneration and living conditions are better for migrants who do skilled or professional work than for those who work on construction sites or do manual labour. All contracts in the Gulf, whether for skilled or unskilled employees from India, are of limited duration. Typically, these are granted for three to four years and may be extended for one or two terms. Unlike migration from Kerala to the United States or Europe, permanent settlement in the Gulf is never possible or intended. In all but a few cases, the families of migrants remain in Kerala and are only seen on home visits every two or three years. Back home, difficulties and hardships of the migration experience are commonly denied in order to produce images of successful upward mobility.[25] Demanding, and often poorly paid labour may euphemistically be termed 'work in a company'. The many Chamakkala men who work on the cleaning staff of the Dubai International Airport, shorten their job description to 'work at the airport', thus alluding to modern, high-status employment. Photographs sent or brought home by migrants commonly show them sitting at an office desk. A noticeable difference in the migrant populations of Chamakkala Muslims and Araya are the relative fractions of countries in which they work. At present, 138 Muslim men from Chamakkala are employed in the Gulf, while an additional 20 men who formerly worked there have resettled in the village. Of these 158 men, 131 (83 per cent) are or have been employed in Dubai. The remaining migrants are (or have been) in other Gulf states, including Saudi Arabia, Abu Dhabi and Oman. Araya Gulf migrants from Chamakkala are not as concentrated on one Gulf state. Of the 44 Araya men, 12 (27 per cent) and 11 (25 per cent) are working in Dubai and Ajuman, respectively.

The predominance of Chamakkala Muslims migrating to Dubai hints at the establishment of social networks with this particular Gulf state. Muslims often benefit from networks related to the mosque. One Muslim Gulf migrant told me: 'People from the mosque moved to the Gulf one by one, like a chain'. In contrast, Hindu migration to the Arabian Gulf is more individualistic and relies, at least initially, far less on networks. Most obtain visas through paid agents. A few Hindus, including some Arayas, have managed to start businesses in Arabian Gulf states. As mentioned above, house members are the benefactors. In the aftermath of the riots following the destruction of the Babari mosque in 1992 many Hindus who had been employed by Muslim-owned companies were laid off and forced to return to India. Many have not succeeded in finding new work, nor have they returned to the Arabian Gulf as migrants.

Signifying West and East

I have introduced the fishermen's conceptualisation of local space, particularly their notion of the *kiḻakkuppuṟaṃ* (the 'outside in the east'; see Figure 2.6) and the *paṭiññāṟụ* ('west') or *paṭiññāṟuppuṟaṃ* ('outside in the east'). These spaces are associated with a variety of meanings, which differ somewhat from one person to another, but altogether they illustrate that the production of locality is a relational process. The west (*paṭiññāṟụ*) is the sea and fishing is the (manual) 'work of the west' (*paṭinjāttụ pani*). Perceptions of the west are largely derived from the work of the west. It is the lineage work (*kula toḻil*), which was done by fathers and forefathers. Fishing is self-determined: in contrast to much land-based agricultural work where labourers may be supervised by landowners at any time, on the sea, i.e. 'in the west', decisions can be made independently of moneylenders or other superiors. Older men often talk about the particular excitement (*āvēśaṃ*), enthusiasm (*jyaraṃ*) or appetite (*vāsana*) for fishing and the linkage (*āśaṃ*) to the seashore as their motivation to become fishermen. On touching the seawater, the fishermen's mind is said to change (inf. *parivarttikkuka*) and becomes more attentive. There is a saying: *mīn kandāl arayan pōle* – 'To see a fish [and get enraged] is like an Araya'. At first, I expected Chamakkala fishermen to disagree with this saying. After all, I thought, it portrays the Araya fishermen (negatively) as short-tempered people. But they did not disagree and rather argued that this excitement is synonymous with an increased perception and concentration and is regarded (positively) as necessary for success in fishing.

In this way the 'west' differs from 'outside in the east' (*kiḻakkuppuṟaṃ*), and, likewise, it is believed that the people of the west must differ from the 'people of the east' (the *kiḻakkāranmar*). The east and its people are not as such negatively valued. Of course, it has always been clear that higher-status people, many of them of immense power, live in the east. As mentioned above, the oppressive effects of this power on the fishermen cannot be ignored and has often been felt painfully in the past. The prohibition for Arayas to walk on public roads and the long-lasting prohibition to enter large temples are further hints at the symbolic subjection of the fishermen by the *kiḻakkāranmar*, who, in the talk of older fishermen, seem to melt into one category.[26] But, on the whole, when older men told me about their experiences with the 'east' and the 'easterners', they seemed either disinterested (perhaps accepting that an 'appropriation of the east' was impossible) or disillusioned.

Thus, while the fishermen of the western coastal belt have been governed by people of the east (*kil̠akkāranmar*), there has been a counter-discourse of the fishermen's independence and moral superiority. It is an inversion of the perception, shared by many of the 'inlanders', that the fisherfolk are backward dwellers on the fringe of an agricultural society.

Figure 2.6: In the kil̠akkuppur̠aṃ *('outside in the east'): the village centre of Chentrappini, on the road from Mumbai (Bombay) to Kochi.*

Notes

1. District, *taluk*, *panchayat* and *ward* are administrative units: Thrissur District is one of Kerala's districts and is further divided into *taluk*-s, which are divided into *panchayat*-s. Edathirutty *panchayat*, in which Chentrappini is situated, is divided into eleven wards.
2. Due to the initiative of a former communist government, every household is given a ration card and is thereby entitled to purchase a defined amount of household supplies (like rice, sugar, kerosene) at subsidised rates at a ration shop. Osella and Osella (2000: 147) note that rice from ration stores is often felt to be of poor quality.
3. Muslim houses are easily recognisable by the names of the householder on the gate or the large panoramic colour posters near the front doors, usually showing the Kaaba and its surroundings in Mecca (Saudi Arabia). A number of Muslims from Chentrappini and Chamakkala have done the *haj* (Arab., pilgrimage) to Mecca, many proudly emphasising the suffix *-haji* in their name. Some even change the writing of the name on the compound's front gate accordingly.

   Among the hints pointing at Hindu residents are the images of Hindu deities on a house's eastern side, ancestral shrines, *citrakūdaṃ*-s (small shrines for *nāga*-s, snake deities) or *brahmarakṣassụ* (shrines for Brahminical spirits residing in the compound), all usually situated in the south-eastern corner of a plot. Since Hindu houses and compounds in Kerala are built or arranged according to a set of fixed rules (e.g. main entrance to the east, kitchen in the north-east, well in the north-east, toilet in the north-west), they may be easily recognised as such (see Moore 1990; Namputhirippaattu 1997). Muslim architects do not feel obliged to follow these rules. However, in Chamakkala there are several Muslim-built houses inhabited by Hindus and a few Hindu-built houses inhabited by Muslims. The plans of Gulf- and North American-inspired villas are often developed by architects residing in Arabian Gulf states.

4. This road was named after Tippu Sultan of Mysore, the son of Hyder Ali of Mysore who, in 1766, invaded Malabar with his army. Malabar remained occupied until 1792; see Dale (1980: 79–91) and Gabriel (1996: 163–237).
5. When I heard references to the north, speakers were mostly referring to Mukkuvar- and Muslim-dominated villages near Chavakkadu and Ponnani. The south is often referred to as the direction from where new fishing technology was introduced, notably Kochi and Allapuzha District.
6. The Malayalam term for inside is *akaṃ*. In the literature the categories *akaṃ* and *puṟaṃ* have been associated with Tamil poetry of the Sangam era (see Ramanujan 1967): *akaṃ* poems are concerned with love and other private emotions, while *puṟaṃ* poems deal with public events such as war. These associations of inside and outside are not linked to the cardinal points. However, as Ramanujan elaborates in the afterword of *The Interior Landscape* (1967), *akaṃ* and *puṟaṃ* are linked to a concept of five landscapes, which are associated with specific sets of commonplaces.
7. The beach is occasionally called the *paṭiññāṟuppuṟaṃ*, the 'outside in the west'.
8. Literature about the Araya caste includes Ananthakrishna Iyer (1909: 261–266), Thurston (1909: VII, 291–298), Klausen (1968: 103–118), Rao Bahadur Govindan Memorial Committee (1973), Vishalakshmi (1973), Mathur (1979) and Padmanabha Menon (1993: 463–476).
9. The name of the compiler of the *Census Report* (M. Sankara Menon) indicates that he was a Nayar and, since the *Jaathi Nirnnaya* was composed in Sanskrit, its author most probably belonged to a high-status caste as well. Thus, inevitably, the caste ranking at which they arrived sees their castes at its top. Fuller (1976: 33–34) arrives at a similar ranking by referring to the older literature but cautions that this order is 'not enshrined in any legal or religious code, nor is it immutable' (ibid.: 33).
10. In the mid-nineteenth century, Francis Day observed the enforcement of the rules of distance pollution, at least with regard to the Mukkuvars (1863: 316). However, since Day claims that the Mukkuvars ranked in status below the 'Chogans' (Izhavas) while the Arayas allegedly claimed to be of a superior status (ibid.: 325) it is not certain whether Arayas were considered by high-status caste members as being as polluting as Mukkuvars.
11. The sacred thread is worn by 'twice-born' Brahmin and Viswakarma males only (Flood 1996: 204).
12. A similar regulation was enforced at the Kodungallur Bhadrakali temple, fifteen kilometres south of Chamakkala. Its annual Bharani festival was opened with the ritual pollution by Araya fishermen, who were not allowed to enter the temple on other days (see Induchoodan 1969: 107–108; Tarabout 1986: 375–377).
13. The word *dīvara* derives from the Sanskrit root *dī* (intellect, intelligence, understanding, mind, idea, thought, brain).
14. Membership in the *dīvarasabha* is restricted to Hindus. Some Muslim fishermen who belong to the same beach (*kaṭavu*) feel excluded from decision-making with regard to matters relating to the entire beach.
15. Altogether, 149 marriages were registered by the *dīvarasabha* in Chamakkala between 1982 and mid-2000. The register is incomplete since some marriages may be registered in the register of the partner's village. Those Arayas who do not want to be associated any more with fishing as the caste occupation may register their marriage elsewhere.

16. The institution of *kadakkoṭi* seems to be identical to that of *ponamban* as described by Padmanabha Menon (1993: 467). He notes that above the village's *ponamban*-s there was a *valia* (big) *aravan* in Cochin. None of the fishermen I interviewed in Chamakkala knew about this institution. The Araya caste was, however, further separated into Arayas and Narusserys. The latter were the barbers and funeral priests of the Arayas but did not intermarry with them until recently. The term *kadakkoṭi* as understood in Chamakkala differs from what Kurien (2000b: 21) writes about northern Kerala. Apparently, there a *kadakodi* is the 'court of the sea', a local institution that decides about rights of access. He claims that this institution was common in north Kerala's Kozhikodu, Kannur and Kasargode Districts.
17. Employment as an IAS (Indian Administrative Service) officer is of high status. One Araya man from Chamakkala works as an IAS officer in higher government employment in Gujarat.
18. The literature about Kerala Muslims includes Thurston (1909: IV, 455–501), Ananthakrishna Iyer (1912: 459–484), Mathur (1978), Dale (1980), Miller (1992), Padmanabha Menon (1993: II, 533–563), Engineer (1995) and Gabriel (1996).
19. Of the distinction of Kerala Muslims into Thangals, Arabis, Malabaris, Puislans and Ossans mentioned by Miller (1992: 253), in Chamakkala explicit reference is made to (high-status) Thangals only.
20. In contrast to the Hanafi Sunnis of northern India, the predominance of the Shafi'ii school in Kerala indicates the South Arabian origin of local Islam.
21. According to Annemarie Schimmel, Mohiuddin Sheikh, alias Abd al-Kaadir (1077 or 1078–1166), is 'probably the most popular saint in the Islamic world' (1975: 247), being associated with a number of miracles. The word 'Mohiuddin' stems from the Arabian Muhyi ud-Diin (lit. 'the reviver of religion'). On Abd al-Kaadir, see Braune (1960). I thank Caroline Osella for drawing my attention to the identity of Mohiuddin Sheikh and Abd al-Kaadir.
22. At the 1999 election for the Indian parliament, the Lok Sabha, about 1100 votes were registered in Chamakkala. Of these 650 (59 per cent) were for the UDF ('United Democratic Front', a coalition of Congress (I), Muslim League and Kerala Congress), 350 (32 per cent) for the LDF ('Left Democratic Front', i.e. the Marxist party CPI(M), the Marxist-Leninist CPI(ML) and some smaller parties) and 100 (9 per cent) for the BJP ('Bharatiya Janata Party', Hindu Nationalist party). At the 1996 *panchayath* election there was a similar outcome in the ward: the UDF, whose candidate was a Muslim, was elected with 49.5 per cent of the votes (592 of 1192). The DLP (Democratic Labour Party) which is associated with the Hindu fisherfolk, is remembered to have been strong in the mid-1980s, but has been weak lately due to internal animosities.
23. Migration from Kerala itself has been going on for more than a century. In conversing with Malayalis, one is commonly told that people from Kerala may be found anywhere on Earth. More than that: 'when the first astronauts landed on the Moon, they found a Malayali there, running a tea-shop' is a popular joke cited by Osella and Osella (2000: 75). Meanwhile there is talk that the planet Mars is inhabited by Malayalis as well.
24. I have not kept similar statistics for Arayas, but I estimate that, compared with the year 2000, there may have been fishermen in perhaps an additional 10 to 15 per cent of Araya households one generation ago. To a certain extent the reduced number of fish vendors may reflect a change in the organisation of fisheries. The introduction of the large *vaḷḷaṃ* boats has considerably affected the catches. The bulk catches of today cannot be marketed by small-scale fish vendors any more, but require the availability of ice for cooling and trucks to transport the catches.
25. These observations parallel those made by Osella and Osella (2000: 76).
26. Nevertheless, it is often emphasised that the east is full of less powerful people as well, especially castes of agricultural labourers (including Pulayar and Vettuva) and tribals; see Uchiyamada (1995).

3 Fishing for Knowledge

I have borrowed this chapter's title from an essay by James Wilson (1990), because it illustrates the thesis I want to develop on the following pages: that fishing in Chamakkala is an endeavour of active knowledge-making. Wilson challenges an assumption that used to be common in theoretical studies of the economy of fishing: that fish are randomly or uniformly distributed in the sea. If this were the case, Wilson notes, the catch rate would depend on the fishing effort and the size of the fish stock only (ibid.: 12). No fisherman in Chamakkala would make this assumption. Most of the economically interesting fish occur in shoals (*pelappụ* or *kūṭṭaṃ*), whose movement follows some seasonal patterns, but these are impossible to predict with such precision as would aid fisheries. The problem of locating fish seems formidable. As Wilson writes, the 'ocean is a very large, complex and rapidly changing environment. No individual fisherman acting alone could hope to acquire the experience necessary to establish the regularity or predictability required for its successful exploitation' (ibid.). If so, the acquisition and exchange of knowledge must be a central task for a fisherman. This is particularly so in catching fish that are pelagic, i.e. near the surface, as is common in Chamakkala. Near the coast the sea bottom is structureless and does not provide an underwater topography that would determine the location of fish. In pelagic fisheries the abundance of species fluctuates in a way that even marine biologists can rarely predict.

The fishermen's participation in such an environment is structured by their craft and gear, and thus we must begin by considering this technology and its recent changes.

Fishing Old and New

Vañci-s

Until the mid-1980s, the *vañci* used to be the dominant boat type in Chamakkala: a dugout canoe for five fishermen. The durability of these boats has been reported to be between fifteen to twenty-five (Mathur 1978: 137) up to more than one-hundred years (Ziener and Rasmussen 1958: 18). The only maintenance required was the application of *cālaney*, an oil made from boiled oil sardines, to its hull. *Vañci*-s are sturdy: capsizing usually does not damage them. They were propelled by rowing (inf. *taṇṭu valikkuka*; from *taṇṭu*, oar, and *valikkuka*, to pull) or, whenever possible, by sailing. Most fishing with *vañci*-s took place at visible distances from the beach (up to about five kilometres). The *vañci*-s were equipped with gill and seine nets. The introduction of nylon nets in the late 1960s appears to have been the main change during this period (PCO/SIFFS 1991a: 4). Cotton nets had been used previously.

Fishing from *vañci*-s was done with active and passive gear. Fishing with active gear means to search actively for shoals of fish or individual fish. Once detected, these are caught, usually with surrounding nets. Fishing with passive gear relies on the surmised presence of fish at a specific location. It means to suspend a net or hooks in the water and wait for the fish to entangle themselves or be caught by the hook. Gill nets, i.e. wall-like nets where the fish become entangled with their gills in the meshes, were common as passive gear. Called *lōpu vala* or *nīṭu vala* in Malayalam (from *nīṭu*, long, and *vala*, net), these nets had different mesh sizes according to the species of fish that was meant to be caught: *cāla vala* (oil sardine net, 28 mm), *ayila vala* (mackerel net, 36 mm) and *sravu vala* (shark net, 80 mm). In operation these nets were spread out in the sea and collected after some time (typically one or two hours). Upon return to the beach the fish were picked from the net (inf. *mīn perkkuka*). During my fieldwork in Chamakkala, gill nets were still used in fishing from small dugout canoes (*ceriya vañci*-s) or small plank-built boats.

Another important active fishing technique made use of the *kolli vala* (lit. 'killer net' from inf. *kolluka*, to kill), a net with a bag-like shape and extended wings.[1] It could be operated as a boat seine from two *vañci*-s, but was apparently more often utilised in cooperation with six or seven other *vañci*-s. In this case, two of them spun the net open, facing a shoal of pelagic fish (usually oil sardines or mackerel). Floaters and sinkers held the net in a shape that resembled a huge mouth (*vāya*) with lips (*cuntu*) at its top and its bottom. Another five or six *vañci*-s were on the shoal's other side. Once the *kolli vala* was brought into a suitable position, the two *vañci*-s holding the net and the other *vañci*-s approached each other. On the latter, people beat the boats with wooden sticks to make a noise and scare (inf. *cāttikkuka*) the fish and chase (inf. *celluka*) them into the net.

In Chamakkala, *vañci*-s were owned by individuals who were themselves fishermen or by fishermen households. There was no clear distinction between the owners of the means of production and the workers. As such, the local fishing economy was non-capitalist. Crews usually included kinsmen, friends and neighbours or other men from the village. It was common for Arayas and Muslims to work jointly in a *vañci*.[2] Due to the *vañci*'s long durability it was often passed on to the next generation. Investments in the craft were irregular and rare, but nets lasted for only two or three years and had to be made or bought regularly.

Fishing with gill nets is possible from a single boat, but the operation of the *kolli vala* required the cooperation of up to fifty fishermen on different boats.[3] This was accomplished by forming *set*-s. Each of these units was a group of eight to ten *vañci*-s whose men owned the net collectively. When the collaborative use of the *kolli vala* was considered to be inappropriate, the *vañci*-s of a *set* operated alone by using gill nets. Prior to the motorisation and introduction of *vaḷḷaṃ*-s in the mid-1980s there had been about twenty-seven *vañci*-s in Chamakkala. These belonged to three *set*-s.

While the *vañci*-s were referred to according to the owner's name, the *set*-s were given specific names.[4] The composition of *vañci*-s in each *set* was semi-permanent. If a conflict could not be settled, a *vañci* changed from one *set* to another. Such changes were possible at short notice, but an individual fisherman was bound to fixed dates if he wanted to change from one crew to another. This was permitted each year on 15 *iṭavaṃ* (Malayalam calendar; about 1 June) and had to be arranged much in advance. The man could only change to the new crew if all of his debts to the owner of the *vañci* that he intended to leave had been paid. This condition could make it very difficult for individual fishermen to change to a new crew. In some cases, a fisherman could be 'bought' out of his contract by shifting his debts to the owner of the vañci to which he changed.

Most of the local fishing trade and fish auctioning was in the hands of a few *tarakan*-s (middlemen), who secured this right by giving cash advances to a team (e.g. for buying a new *vañci* or nets). The *tarakan* did the auctioning of the catch and received the *taraku* ('auction fee') for this work, usually 5 to 7 per cent of the money received for the bulk catch. The loan given by the *tarakan* to secure the right to auction the fish had to be repaid separately, including interest. The right of auctioning was held by the *tarakan* until the loan had been repaid. Middlemen were usually less interested in quick repayments and rather wished to prolong their right to auction the catch. Given the observed fluctuations in the catch, the personal relationship with the *tarakan* was risk-minimising for the fishermen, but was often exploitative. Unlike in south Kerala, where female household members market the fish that has been caught by male household members (see Platteau 1984: 72; Nayak 1993: 70ff.), women in Chamakkala were not involved in fish vending. The fish was auctioned to head-load and cycle-load traders and, especially during the south-west monsoon, to larger merchants who came to the beach with a truck.

The division of the money obtained from auctioning the catch was done after subtracting the *taraku*. One-fourth (25 per cent) of the money went to the boat owner and one-fifth (20 per cent) to the so-called *panku kaśụ* account. The remaining 55 per cent was shared equally among the crew members. Thus, if a five-man *vañci* got 1,000 rupees for their catch (after subtraction of the *tarakụ*), each fisherman was paid 110 rupees. The division could be more complicated, e.g. if the boat was owned by one man and the nets were jointly owned by the crew members, as was sometimes the case. The *pankụkaśụ* money was kept by the boat owner as the crew's shared savings account. It was saved over the year until 15 *iṭavaṃ* (approximately 1st of June; typically the onset of the south-west monsoon) and then divided into six shares: one each for the boat owner and the five crew members. If a crew member was in urgent need, a loan could be made from the *pankụkaśụ* prior to the annual division. The *pankụ kaśụ* was a financial reserve for the lean season and was meant to dampen extreme fluctuations in the income. With the introduction of the *vaḷḷaṃ*-s, *pankụkaśụ* accounts were largely abandoned.

Table 3.1 Comparison of the *vañci*, the dominant boat design until about 1980, and the *vaḷḷaṃ*, dominant during the time of fieldwork in 1999–2000.

Craft	*vañci*	*vaḷḷaṃ*
Type of craft	dugout canoe	plank-built boat
Size (length)	about 7 metres	up to 21 metres
Durability	15 to 100 years (see text)	3 to 5 years
Propulsion	rowing	outboard motor, up to three at 40 hp each
Crew	5 men	30 to 35 men
Gear	gill nets (length up to 200 metres; about 20 kg) *kolli* ('killer') net	ring seine nets weight: 700 to 800 kg size: up to 500 metres x 60 metres
Ownership of craft and gear	single men or shareholder	10 to 30 shareholders
Fishing technique	*passive*: laying out the gill net and hauling it on board after some time *active*: watching for *lakṣyaṃ*-s and then trying to chase fish into the net by noise	*active:* watching for *lakṣyaṃ*-s (visual signs) of shoals, surrounding shoal rapidly, pulling net together at bottom and hauling it into the *vaḷḷaṃ*
Marketing of fish	auctioned and sold to local small-scale vendors and merchants; majority of catch sold in the District	bulk catches auctioned and sold to large-scale traders; requires trucks, cooling facilities and marketing structure

Figure 3.1: Pulling a vala vaḷḷaṃ, *a large plank-built boat, on shore.*

Vaḷḷaṃ Units

In 1999–2000 large boats, called *taṅṅụ vaḷḷaṃ* or simply *vallam*, were the dominant fishing craft in Chamakkala, dominant in terms of catch and number of fishermen involved (Figure 3.1). Table 3.1 compares their essential features with those of the *vañci*-s. The *vaḷḷaṃ*-s are suited for the catch of fast-moving, pelagic shoaling fish in calm weather conditions. Operating with encircling nets, these boats are made of wooden planks (sing. *palaka*), which are attached to a wooden frame. The *vaḷḷaṃ* is probably an Arabian design (Hornell 1920: 155). The boats are up to twenty metres long and are motorised. Two or three outboard motors of up to forty horsepower each are attached to their hull. The *vaḷḷaṃ*-s are much less durable than *vañci*-s and require continual efforts of maintenance. They are replaced every three or four years. As such, expenses for their craft are always on the fishermen's bill.

About thirty to fifty men are involved in each *vaḷḷaṃ* unit. This number includes the shareholders, who jointly own craft and gear, and a varying number of 'freelance' workers, who join the team for any length of time, from a single fishing trip to many years. Fishing with *vaḷḷaṃ*-s makes use of encircling ring-seine nets (called *taṅṅụ vala* from inf. *taṅṅuka*, to float, and *vala*, net, or *ring vala*, from the English word *ring*). Gill nets and lines are not employed. The use of ring seines is an active method of fishing. In operation, a shoal of fish is detected by sight. Then the net is laid around it swiftly, closed and then pulled on board by hand or by means of a motorised winch. In 1999–2000, two kinds of ring seines were common: the *ring vala* (mesh size: 18–22 mm), used to catch sardines, mackerel and prawns, and the *cūta vala* (mesh size: 10–12 mm), used to catch white sardines (*cūta*). Fishing with the *ring vala* requires the participation of at least twenty-five men, fishing with the *cūta vala* about fifteen to twenty.

The prevalence of *vaḷḷaṃ* units in Chamakkala reflects a trend that has been noticed throughout central and northern Kerala. According to a survey conducted by the South Indian Federation of Fishermen's Societies (SIFFS), 2,271 ring-seine nets were operated by 1,636 *vaḷḷaṃ* units in Kerala in 1997–98. The number has not changed much in comparison with a previous census conducted in 1991 (2,259 ring seine nets operated by 1,741 *vaḷḷaṃ* units; SIFFS 1992: 31, 35). In the 1990s these units accounted for about half of the fish landings in Kerala (D'Cruz 1998: 34). Since the first units began operating in 1985, these numbers suggest that a rapid spread of this technology must have taken place in the second half of the 1980s. The origins of this 'boom' and the processes involved therein are best understood by considering a *vaḷḷaṃ* unit in more detail. Even though I shall introduce only one such unit from Chamakkala, named Ponnonam (lit. 'golden Onam'[5]) as it was constituted in early 2000, I have taken my information from other local *vaḷḷaṃ* units as well.

Founded in 1988, Ponnonam was the largest and economically most succesful *vaḷḷaṃ* unit in Chamakkala in 1999–2000. At that time its craft included four *vaḷḷaṃ*-s (the largest with a motorised winch) and a *muṭuveṭṭi*, eight outboard motors, two *ring vala*-s and one *cūta vala*. There were fifteen shareholders and fifty-three men who work, more or less regularly, in the unit. The debt in early 2000 was about 200,000 rupees, most of it borrowed from the *tarakan*.

A look at the other *vaḷḷaṃ* units in Chamakkala reveals differences in organisation: Svayambhu, the second largest unit, smaller than Ponnonam in terms of equipment, is owned by as many as thirty-two shareholders. Ambadi and Eran, the smaller *vaḷḷaṃ* units, are owned by individuals. Both are Gulf returnees from the same lineage. While the Ambadi team operates with gill nets from *muṭuveṭṭi*-s most of the year, the Eran *vaḷḷaṃ* operates during the south-west monsoon season only. Its crew is recruited among fishermen in Kerala's Kollam district, about 200 kilometres to the south.

The introduction of the *vaḷḷaṃ* in central Kerala must be seen in the light of some profound changes that have taken place in the state's fisheries since the 1960s, notably the discovery of prawn resources, the emergence of an extensive fleet of trawlers and the impact of severe competition on the artisanal fishermen (see Chapter 6).

When I asked fishermen in Chamakkala about the origin of the *vaḷḷaṃ* units, the usual response was that they had come 'from the south', i.e. from

Kerala's Kochi, Kollam or Allapuzha Districts. In Thrissur District, the *vaḷḷaṃ*-s were apparently first seen during the south-west monsoon in 1983.[6] Initially, there was a strong opposition against these foreign (*vidēśi*) boats. They were not allowed to land at local beaches. But soon this ban was lifted. The *vaḷḷaṃ*-s were accommodated at the beaches and food was sold to their crews. Some local fishermen joined the units as guest workers and got first insights into the new technology. Nevertheless, the *vaḷḷaṃ*-s were feared as a threat to the established local fishing economy. It had been noticed that the *vaḷḷaṃ*-s could bring large catches of oil sardines and mackerel to the shore within a short time. Even though a number of local *vañci*-s had been equipped with 8 hp outboard motors in the early 1980s, they could carry only about one-tenth or one-twentieth of a *vaḷḷaṃ*'s catch and arrived at the beach long after the catch of the *vaḷḷaṃ*-s had been auctioned. In the meantime, the auction rates had often fallen considerably.

Matsyafed, then a recently founded government apex organisation of fishermen's co-operatives, began to provide subsidised 8 hp outboard motors for the *vañci*-s in 1984. In Chamakkala, this programme was a failure, since many people did not trust the government loan schemes. Some men believed that these motors were an idle investment, since *vaḷḷaṃ*-s would soon become the principal craft anyway. Others believed that both *vaḷḷaṃ*-s and outboard motors were just two costly and short-lived fashions and would soon disappear.

The first *vaḷḷaṃ* unit in Thrissur District began operating in Azhikodu in 1984. A year later several units began in Perinjanam, Kaipamangalam and Nattika. Initially, the catches of these units were impressive. In particular, there were good catches of prawns in the mid-1980s. These were regarded as a strong incentive to form more *vaḷḷaṃ* units. By 1987 or 1988 there were *vaḷḷaṃ* units at every *kaṭavu* in Thrissur District. The *vañci set*-s dissolved and fishing with the boat seine net (*kolli vala*) was abandoned. Several *vañci* teams stopped operating one or two years after the *vaḷḷaṃ*-s became commonplace. A few small teams continued to fish with gill nets from motorised *vañci*-s until the mid-1990s.

The introduction of the *vaḷḷaṃ*-s quickly transformed long-established relations between fishermen, boat owners and middlemen. The investment necessary to form a *vaḷḷaṃ* team is similar to the expenses for a trawling boat (Kurien and Vijayan 1995: 1781). In the late 1980s it amounted to at least 300,000 rupees, a sum that was difficult to procure even for *tarakan*-s. At this initial stage, the state cooperative Matsyafed was crucial in providing loans directly for registered groups of fishermen. It was the aim of Matsyafed to free fishermen from the ties of the *tarakan*-s. The loans were meant to bypass their influence. In order to avoid the concentration of money in the hands of a few men (as had happened with the 'society boats'), a minimum number of eighteen active fishermen was required to

Figure 3.2: A ring-seine net, used on a vaḷḷaṃ, *is brought to the shore.*

be eligible for a loan from Matsyafed. But not all units were formed with its help.

In founding the Ponnonam unit, eighteen men got together to apply to Matsyafed for a loan of 200,000 rupees. Most of these men had worked in *vañci* teams before their dissolution. The assembly of the shareholders was driven by kinship ties and personal friendship. The loan was granted and additional loans were secured from several private moneylenders and from a *tarakan*, who thus secured the right to auction the catch. A second-hand *vaḷḷaṃ* and a 25 hp motor were bought from Kollam District. The initial ring-seine net was assembled from 300 kilograms of net subsidised by Matsyafed and 50 kilograms paid for with the money of a private loan. This net was about 190 *maru̧* long and 20 *maru̧* deep.

Men who had been bonded to specific *vañci*-s as workers via long-term contracts now felt themselves free to be the co-owners of craft and gear. The initial resentment against this technology did not trigger attempts at developing a local alternative to the foreign technology. Rather, the new technology had come to Thrissur District and came to stay.

Comparing the initial teams with their condition in 1999–2000 one notices that in both cases the number of shareholders has slightly decreased while the sizes of boats and nets and the power of the outboard motors have much increased. For example, between 1988 and 2000 the length of the major *vaḷḷaṃ* of the Ponnonam unit increased from 14 to 20 metres, its propulsion from one 25 hp engine to three 40 hp engines, and the net size (length) from 190 *maru̦* to 325 *maru̦*. The latter corresponds to a weight of about 1,000 kilograms (Figure 3.2).[7] These changes have been incremental since the unit began operating. Similar increases in the size of craft and gear have been noticed throughout Kerala (Rajan 1993; D'Cruz 1998; SIFFS 1999; Kurien 2000c). Another local unit, beset with organisational and financial problems, could not adequately participate in this competitive development and stopped operating.[8] While an increase in boat and net sizes and engine power has been noticed throughout Kerala, this development has been particularly pronounced in Thrissur district, where smooth surf conditions and the structure of the sea bottom permit the use of large nets (see D'Cruz 1998: 13).

Organisation of a *Vaḷḷaṃ* Unit

Let us look in a little more detail at the crew composition of the Ponnonam unit from Chamakkala. As mentioned above, about fifty-three men are associated with this unit, fifteen of whom are its shareholders. In autumn 1999, I conducted a survey of these men, enquiring about their family background and economic situation, their work in the unit and their previous experience in fishing. The ages of crew members varied between seventeen years and sixty-eight years. The youngest shareholder was thirty-seven years old. Most of the twenty-seven men at or above this age have had previous experience of fishing with *vañci*-s. Several men had left fishing for periods between a few weeks and more than ten years, doing cleaning work in a tea shop in Madras or construction work in Arabian Gulf states.

Younger men usually lack the experience of fishing on *vañci*-s. About eighteen men with ages between seventeen and thirty-two years worked on Ponnonam. Having typically left school before graduating with SSLC, many of them tried to find work inland, the *kiḻakkuppuṟaṃ*, before beginning to work on the *vaḷḷaṃ*. Some were doing irregular work on the side, e.g. as painters or money collectors for a chitty fund (savings club).

Most men described their work on the *vaḷḷaṃ* as *vala vali[kkuka]* ('pulling the net'). Like launching the large boats into the sea to embark on a fishing trip and returning them to the beach thereafter, this is unskilled, manual work (*pani*). As previously mentioned by Kurien and Vijayan (1995: 1783), few skilled crew members are needed in operating a ring-seine unit. An experienced skipper (*aryakkāran*) decides whether to embark on a fishing trip. He guides the craft, spots the fish and directs the fishing operation. There is also an engine coordinator (*sranku̦*) and the engine and winch operators. If the membership of a unit is defined by those men who receive a share of the financial income, the ritual specialist (*mantravādi*) and even some deities must be included.

Several men may coordinate the operation of the engines (and thus be the *sranku̧*), work that essentially consists of translating the hand signs of the *aryakkāran*, who stands at the helm of the *vaḷḷaṃ* while it is moving, into commands to the engine operators. These engine operators are always young men who are versatile and enthusiastic about the operation of the outboard motors. They usually have some skills in engine maintenance and know how to do simple engine repair jobs. Another two men are responsible for the operation of the winch, driven by a diesel truck engine fitted into the *vaḷḷaṃ*. The winch is meant to close the bottom part of the net and pull it near the *vaḷḷaṃ* to complete the fishing operation. Several young men are knowledgeable about the operation of the outboard motors, but many of them are neither skilled nor interested in mending the nets. This work is done by older men. As such there is a division of labour between the more traditional and recent kinds of fishing-related work. Net repairs are usually done at the beach.

While some initial shareholders sold their shares, others bought some. But altogether the number of shareholders has decreased from eighteen to fifteen. Such a decline in the number of shareholders has been observed for other units as well (see Kurien and Vijayan 1995: 1783). While Kurien and Vijayan attribute this to conflict in the groups, it may also mirror the fact that some men appear to be only ostensible owners of their equipment, much of the de facto ownership actually resting with the *tarakan*.

There are few fluctuations in the group of men who work on the Ponnonam *vaḷḷaṃ*. Most have been associated with the team for many years, but there are few obstacles for an Araya man from Chamakkala to join the unit as a worker for a day or longer. It is generally agreed that a man's request to work in a *vaḷḷaṃ* unit cannot be declined. Everybody who comes to the beach in the morning before launching the *vaḷḷaṃ* is counted as a worker for that day and is entitled to a share in the division of the money, even though he may not have joined the crew in the fishing trip and may not have done any work on that day.[9] If there is a surplus in the available labour force, the income is spread among all workers. According to Kurien and Vijayan (1995) this custom stems from Allapuzha District and spread along the Kerala coast with the *vaḷḷaṃ*-s in the late 1980s. They refer to it as *karanila* (from *kara*, shore, and *nila*, state, position; the term is not explicitly in use in Chamakkala) and call it a mechanism of income spreading that serves as a social security system (Kurien and Vijayan 1995: 1784). Other authors have referred to this custom as 'unemployment in disguise' (PCO/SIFFS 1991a; Rajan 1993). Kurien and Vijayan emphasise that because of the small number of shareholders the *karanila* system is necessary to supply an adequate supply of workers for the operation of the ring-seine units. If the *karanila* share were not paid, they argue, more fishermen would attempt to leave fishing permanently and endanger the continuation of the present mode of operation (Kurien and Vijayan 1995: 1784). One might add that this custom may be suited to adjusting to seasonal variations in the demand for workers. As it is practised now, the custom makes it easier for Chamakkala men to shift between fishing, the work of the west, and working in the east. It stands in marked contrast to the bonded contracts that prevailed in the operation of the *vañci*-s and made it impossible for fishermen to be flexible in such a way. Besides these apparent economic benefits, the necessity for allowing anybody to participate is sometimes put down to the danger of the team being affected by the evil eye (*kariṃ kaṇṇụ*) of the man whose offer to work was turned down. Similarly, in any cash distribution all persons present, including children, are entitled to at least a small share in the money. This is also meant to prevent effects of the evil eye.

The income from fishing is distributed among the workers according to fixed rules which differ from team to team. The mode practised by the Ponnonam unit is fairly typical. After auctioning the catch the *tarakan* deducts for himself the sales commission (*tarakụ*), usually 5 to 7 per cent. The remaining money is split into two equal fractions. One half is distributed to all the workers of the day, the other half is used to meet the expenses for fuel, engines, repayment of loans and the food that is provided for all workers (usually in a simple restaurant or tea stall). A shareholder working on a certain day receives one share as a worker and an additional, usually much smaller, share from the shareholders' fraction (this fraction is usually saved in a joint account, the money being distributed only once or twice a year). Not all workers receive the same share. Being particularly skilled, the *aryakkāran* receives one and a half or two shares. Every team is associated with a ritual specialist (*mantravādi*), who conducts astrological consultations for the unit, does *pūja*-s in its name and regularly charges the boat's *yantraṃ* (ritual diagram engraved on a copper plate) with *mantraṃ*-s (magical spells). This person often receives two shares, one for his ritual work and the other one for the deity invoked by the rituals, which he performs in his temple. A separate account is kept for offerings (*vaḷipaṭụ*) that are regularly brought to outside temples.

Two Fishing Trips on the Ponnonam *Vaḷḷaṃ*

Let us now consider two fishing trips that were undertaken by the Ponnonam *vaḷḷaṃ* unit during my fieldwork in Chamakkala. While sharing many features, these two trips illustrate the variety of decision-making and the changing fisheries environment that confronts Chamakkala fishermen.

I begin with a description of a fishing trip conducted on 12 November 1999. This was towards the end of the north-east monsoon (*tulavarśaṃ*). Calm weather conditions prevailed on that day. During the north-west monsoon, evening thunderstorms and rain showers are common, but in mid-November these were lessening. The *vaḷḷaṃ* had gone fishing almost daily over the last two weeks. In early November daily catches were satisfactory and had amounted to 10,000 to 30,000 rupees' worth. In the days prior to the trip on 12 November, catches had ranged from 2,000 to 5,000 rupees, which is less than the expenses for kerosene and oil. Following the fishermen's parlance, I shall refer to the main boat from where the fishing is conducted as the *vala vaḷḷaṃ* ('net boat'). The carrier boat, on which the catch is brought to the auction place, will be referred to as the *dinky vaḷḷaṃ*.

On this day, thirty-four men are involved in the fishing trip on the boats, a few more men help with the launching and landing of the boats on the beach. Joseph is the *aryakkāran* (skipper). Abdu Puzhamkaraillathu, sixty-two years old, is the *sranku̥* (engine coordinator of the outboard motors). Kuttan Arayangattu, thirty years old, operates the winch. Sijeesh Kovil Tekku Valappil, seventeen years old, today the youngest man on board, will be the *caṭṭukkāran*, the 'jumping man' (see below).

At 5.30 a.m. all men meet on the beach near the *bhanḍāraṃ* box. Joseph and some other middle-aged fishermen discuss the weather and surf conditions. The equipment is brought to the beach from the unit's shed. The outboard motors, oilcans, kerosene cans and the battery of the winch are brought to the *dinky vaḷḷaṃ*. The *vala vaḷḷaṃ*, the main boat in today's fishing operation, had been anchored in the sea for the last week or so. The fishing operation begins by loading oil and kerosene onto the *dinky vaḷḷaṃ* to take them to the *vala vaḷḷaṃ*. Before launching the *dinky vaḷḷaṃ* into the sea, all engines are briefly tested. At 5.55 it starts out to the sea with thirty-one men. Three men are starting out to the sea separately on a *muṭuveṭṭi*. At 6.00 the *dinky vaḷḷaṃ* has reached the *vala vaḷḷaṃ*. Twenty-eight men shift to the *vala vaḷḷaṃ*; three remain on the *dinky vaḷḷaṃ*. The men on the *vala vaḷḷaṃ* remove a large plastic sheet cover from the boat, remove seagull droppings, fill the tanks with kerosene, start the motors and briefly test the functioning of the winch. Then the men assume their places for the operation of the boat. Being the *aryakkāran*, Joseph stands in the front (*aniyaṃ*) whenever the boat is moving. Abdu, the *sranku̥*, sits in the rear end (*amaraṃ*). On the front and the back of the *vala vaḷḷaṃ*, a small *pūja* is performed with *karpūra ārati* (camphor flame). Incense is put in front of the *yantraṃ*, which has been brought onto the *vala vaḷḷaṃ* for the fishing trip and which had been previously charged by Balakrishnan, then the unit's *mantravādi*.

At 6.10 a.m. the boats begin to move westwards, guided by Joseph's hand signals to Abdu who operates one engine himself and directs the

Figure 3.3: Encircling a shoal of oil sardines.

other engine drivers. In raising his right arm swiftly a few times every now and then, he signals to raise the speed or keep going at high speed. When he holds his right hand waist-high, palm facing down, and moves it up and down gently he signals to slow down.

In moving, the *dinky vaḷḷaṃ* and the *muṭuveṭṭi* stay to the left and right of the *vala vaḷḷaṃ*, respectively. Men from all three boats watch out for the signs of fish, the searched area of the sea thus being enlarged. Further to the west, two other *vaḷḷaṃ* units are seen. Suddenly one of their boats moves in a circle. When this movement is completed a cloud of exhaust fumes rises from the boat. Apparently that team had made a catch, the exhaust fumes indicating the start of the winch used for pulling the net aboard. Joseph directs the boats towards these *vaḷḷaṃ*-s in order to inspect their catch. It turns out to be a meagre catch of oil sardines (*cāla*). Joseph directs the Ponnonam *vaḷḷaṃ's* search for fish in the vicinity. He is joined by two men who stand in the helm. They identify a shoal (*pelappu̦*) of fish by ripples on the water surface. The *vaḷḷaṃ* slows down for the men to have a closer look without disturbing it, but, after discussing it, Joseph and Kuttimon deem it not worthy of being caught. The time is now 6.40 a.m. Speed is increased, and soon another shoal is spotted. Like the first one, it is regarded as too small. On arriving at a third shoal, the speed is lowered. After another look and a short exchange with Kuttimon and two other men in the *aniyaṃ* (front) of the *vaḷḷaṃ*, Joseph decides to catch it. The shoal is approached carefully. When it is seen on the left side, Joseph slowly raises his left arm and then turns it down quickly. This is the signal for the *cāṭṭukkāran* to jump into the water with one end of the net. The speed of the *vaḷḷaṃ* is immediately increased and the shoal is quickly surrounded in an anti-clockwise motion (Figure 3.3). Throughout the operation, directions are given by the *aryakkāran* (Joseph) to the *srānku̦* (Abdu). While the *ring vala* net runs into the water, the *valakkāran*-s (lit. 'net men', the workers) have to pay attention so that it runs smoothly over the boat's side (*viḷḷu̦*). At this moment the winch, with a diesel truck engine, is started. Three minutes later, the shoal has been encircled. The *vaḷḷaṃ* returns to the *cāṭṭukkāran* and takes him on board. Both ends of the net's bottom rope (*kamba*) are now being pulled by the winch and thus the bottom of the net is closed. Once completed, the winch motor is stopped. Meanwhile the men stand along the entire left side of the *vaḷḷaṃ* and start to pull (inf. *valikkuka*) the net on board. Some sardines jump across the net. To prevent the escape of fish under the bottom of the *vaḷḷaṃ*, the *cāṭṭukkāran* and another young man keep jumping into the water in the middle of the boat, in order to scare the fish back into the net. While pulling the net, the men sing work songs, most ending in a refrain '... Ailassa!'[10] When the net is almost completely on board, the largest part of the shoal is in the remainder of the net. It is a small catch of sardines. The *muṭuveṭṭi* is sufficient to bring it to the auction place on the seashore (today this is at Companykatavu, south of Chamakkala). The catch is slowly put into the *muṭuveṭṭi*, which soon takes off.

After a short break, the net is inspected and cleaned in the sea. Some small holes are mended, and then the net is put back into the rear half (*amaraṃ*) of the *vala vaḷḷaṃ*. While this work continues, the *vala vaḷḷaṃ* and *dinky vaḷḷaṃ* start to move south, parallel to the coast, heading for another *vaḷḷaṃ* unit, which is seen securing a catch. At 8.00, the boats move further south-west slowly away from the land in some wide turns. While Joseph, Thambi and a few other men watch the surface of the sea carefully, looking for the signs (sing. *lakṣyaṃ*) of fish, most of the workers are talking, sleeping or telling jokes. The boats move further to the west. At about 9.00, land is out of sight. Some shoals are spotted and the biggest of them is approached. Joseph gives the signal for the *cāṭṭukkāran* to jump into the water. The procedure described above ensues, but this time a lot of fish escape the net. And what is worse is that, while the net is being pulled on board, dolphins (*kaṭalēṭi*) are approaching, trying to feed on the catch and bite the net. There is a lot of rushed excitement and shouting. The *muṭuveṭṭi*, which has returned from delivering the first catch to the auction place, moves around the net, one man beating the water surface with an oar in order to scare the dolphins away. The resulting catch is disappointingly small. The *muṭuveṭṭi* sets out to deliver it to the auction place at Companykatavu.

At 9.30 the two *vaḷḷaṃ*-s move east until the land is back in sight. Then, careful mending of the net begins: the dolphins have caused some big holes, which are closed (*vala aṭṭakkuka*). At 10.30 this work is finished. The *muṭuveṭṭi* has returned from the auction place. Now the *vala vaḷḷaṃ* begins to move towards the west-north-west into the sea. As before, the *dinky vaḷḷaṃ* and the *muṭuveṭṭi* run parallel on both sides. While Joseph and a few other men in the front carefully watch the sea, most men on board are sleeping. By now the heat of the day has much increased, only one other *vaḷḷaṃ* is seen searching for fish in the distance, most of the boats have already returned to the shore. The sea looks dead. There are no signs of any shoal, not even a single dolphin. Joseph directs the boat to move westward. After forty-five minutes of not seeing any shoal, Joseph gives the signal to return. At 12.30 p.m. we arrive in Chamakkala. The *vala vaḷḷaṃ* is anchored off the shore and is covered with a plastic sheet. All men return to the beach on the *dinky vaḷḷaṃ* and the *muṭuveṭṭi*. A few of the younger men remove the outboard motors from these two boats, clean them with fresh water and return them to the shed. The oil and kerosene cans and the battery of the winch are returned as well. The *yantraṃ* is returned to Balakrishnan's temple. Once this work is finished, the money is distributed. Today's catch was auctioned for a total of 4,500 rupees. Each worker is given 50 rupees (about 1.30 euro). Most men join Joseph to take a

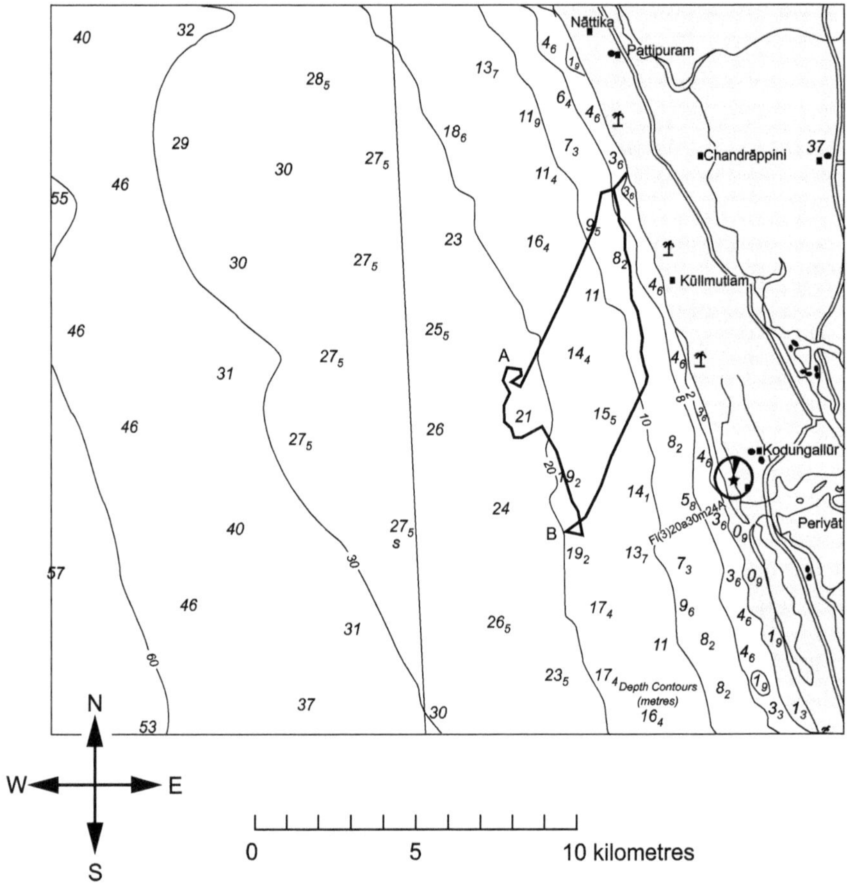

Figure 3.4: Route of the Ponnonam vaḷḷaṃ *on 1 February 2000 as measured with a portable GPS satellite receiver, shown on Nautical Map DHI 1565 (Deutsches Hydrographisches Institut, Hamburg). The depth is in metres. A: first ring-seine fishing operation (8.49 a.m. to 9.56 a.m.). B: second ring-seine fishing operation (11.28 a.m. to 1.00 p.m.).*

meal and a tea at the tea-shop at Three Cross. The expenses are paid from the shareholders' account. Afterwards the men go home and have a wash. Some of them meet again later on the beach near the *bhanḍāraṃ* box; others go to Chentrappini Centre or do work at home.

A trip that took place on 1 February 2000 differed from the other in some important respects, including the decision-making of where to fish in the presence of many trawlers in the sea. This day's *aryakkāran* is Abdu Puzhamkaraillathu, who was the *srankụ* on the trip described above, and the *srānkụ* is Anthony, Joseph's father's younger brother's son (FyBS). The remainder of the team is much the same as in November. On the previous day, the team had caught sardines worth 15,000 rupees 20 kilometres south of Chamakkala. I shall be measuring the boats' movements in the sea by using a portable GPS satellite receiver (Figure 3.4).

At 6.00 a.m. Abdu and some other men are gathered at the beach. They seem undecided. There is a slight breeze from the east, but the sea is calm. The sky is clear. Abdu and some other men talk about whether the sea would be hot (*cūṭu*) or cold (*tanuppu*) and in which way the wind would affect the sea's 'heat' (wind from the east is generally believed to heat the sea). Abdu waits. At 7.30 the Sudarsanam *vaḷḷaṃ* is seen going to sea at the neighbouring *kaṭavu* (Companykatavu). Sudarsanam is known as a successful *vaḷḷaṃ* unit. Abdu gives the command to enter the boats. Thereupon some other *vaḷḷaṃ*-s are seen being launched as well. A large number of trawlers are seen in the sea, some are fishing at a distance of only 500 metres from the beach (according to the Kerala Marine Fisheries Resource Act (KMFRA), trawling is forbidden within ten kilometres of the coast). This is a common sight and none of the fishermen seems particularly concerned.

On this day, the *muṭuveṭṭi* is not used. There are twenty-one men in the *vala vaḷḷaṃ* and four men in the *dinky vaḷḷaṃ*. As in November, the *vala vaḷḷaṃ* has been anchored in the sea approximately 200 metres off the coast. After launching the *dinky vaḷḷaṃ*, the *vala vaḷḷaṃ* is reached at 8.05. It is cleaned, the engines are tested and a small *pūja* is performed. Twelve minutes later, Abdu has taken the helm and directs the *vaḷḷaṃ*-s to move to the south-west. After a few minutes we are surrounded by at least fifty trawling boats. Some of them are operating their trawl nets. At 8.40 land is barely visible any more. Abdu directs the boat to slow down and come to stand still. At a distance, two seagulls (*kaṭal kākka*, lit. 'sea crow') are seen splashing into the water. Getting closer, ripples on the surface hint at the motion of a small shoal of fish. Three minutes later the shoal has disappeared. Abdu decides to turn around and move towards the east. No trawlers or other *vaḷḷaṃ*-s are in our immediate vicinity now. After a few hundred metres we stop again, everybody scans the sea for signs (sing. *lakṣyaṃ*) of fish. Then Abdu directs the crew to turn to the north-west and increase the speed. On seeing a shoal, he gives directions to approach it and then gives the command for the *caṭṭukkāran* to jump into the sea with one end of the net (at 8.49). The same fishing procedure as described above ensues. After about forty-five minutes, the catch is hauled onto the *dinky vaḷḷaṃ*. It is a mix of white sardines (*cūta*), small prawns (*cemmīn*), some shells from the sea bottom and a plastic bag. A sea snake that was caught is thrown back into the sea. This catch is too small to be worth being brought to the auction place. The *dinky vaḷḷaṃ* of another unit comes to take a look at our catch. At 9.56 the net is back in the rear of the *vaḷḷaṃ*. Some holes in it are mended. When Abdu gives the signal to proceed, one outboard motor fails. It is fixed after thirty minutes of repair work.

Figure 3.5: Landing a catch of oil sardines at the auction place.

Meanwhile, both *vaḷḷaṃ*-s are surrounded by many trawling boats, which are seen moving from north to south. After the engine has been fixed, both *vaḷḷaṃ*-s roughly follow the trawlers in moving towards the south-west, speeding up to move in front of them (Abdu explains that he anticipates the fish to be chased by the trawler's noise). When a few seagulls are seen over one spot in the water, Abdu gives the signal to slow down and approach it. At first, the men watch for fish in vain. After proceeding towards the shore, a small shoal of white sardines (*cūta*) is seen. It is now 10.55. Inside this shoal a bustling motion is noticeable, as if waves were propagating through it, one side of it being loosely defined, the other side like a sharp edge. Whenever the wave of motion arrives at this edge, some fish jump out of the water and are visible. The men watch it with tense concentration. Everybody is on his position now, but Abdu seems undecided. The trawlers are approaching from the north, a *vaḷḷaṃ* with an empty *dinky* passes by from the south (a hint at the lack of fish in that direction). Abdu instructs the *cāṭṭukkāran* to get into the position to jump. But then he decides not to catch this shoal but to proceed further. After moving towards the west and passing two groups of trawlers, a shoal of small sardines (*kuṭṭicāla*) is discovered. After discussing this shoal with two other fishermen at the helm, Abdu decides to catch it. At 11.28 the *cāṭṭukkāran* jumps into the water. The fishing procedure described above ensues. The catch is taken to the auction place at Companykatavu (Figure 3.5). At 13.30 the *vaḷḷaṃ*-s start to return to Chamakkala. Some mending work is done along the way. At 15.00 the *vala vaḷḷaṃ* is anchored in the near-shore water off Chamakkala. At 15.30 the *dinky vaḷḷaṃ* is landed on the beach. The activities at the beach are much the same as in the trip described above. Today's catch was auctioned for 4,000 rupees. Each man receives 50 rupees (about 1.30 euro).

Decision-making and its Constraints

Some key elements of decision-making can be deduced from these descriptions. First, an initial aspiration level has to be surpassed to generate interest to embark on a fishing trip. The primary driving factor is the market auction rate for the catch. This is known to vary with the seasons as well as over periods of weeks, days and hours. Secondly, a fishing trip can commence and continue only when weather and surf conditions are favourable. Thirdly, when at sea the *aryakkāran*'s decisions are governed by watching the sea surface and bird behaviour, as well as making inferences from the motions of other boats. Ideally, several shoals, perhaps of different species, are encountered during a trip and the *aryakkāran* has to decide which one to catch. Visual signs (*lakṣyaṃ*-s) lead him to decide whether or not to slow down and throw the net. He knows that this decision commits the team to at least one hour of work. Since shoals move and disperse rapidly, he rarely attempts to return to a shoal that has been inspected previously. Thus, the rule in capturing seems to 'take the last' when a promising or at least minimal aspiration level is reached. Further constraints are the rising cumulative cost of kerosene in the course of the trip as well as the typically falling auction prices in the course of days on which several teams operate.

In these lines I have described the fishing activity with the vocabulary of the 'adaptive toolbox' of bounded rationality, an approach to everyday decision-making developed by cognitive psychologists that emphasises the perennial presence of constraints in making choices (see Gigerenzer 2002). Expressed in these terms it becomes obvious that the number of constraints has increased with the introduction of the *vaḷḷaṃ*-s and the growing importance of trans-local fish markets.

What struck me most during these trips was the extent to which external information was relevant in decision-making. I had expected, perhaps naively, that decisions would be made overwhelmingly or even solely upon the judgement of local climatic and marine conditions. When these conditions were considered before embarking on a fishing trip, as in Abdu's pondering about the 'heat' or 'coldness' of the sea, they alone did not lead to the decision he reached eventually. Even though specific conditions are judged as beneficial for fishing, fishermen rely less on these than on inferences drawn from watching the movements of other boats or taking a look at their catch, hearsay or mobile phone calls from a fish trader. One man told me that his *vaḷḷaṃ* would go out fishing on the next day if there were a southern current (which he associated with nutrient-rich water attractive for fish). I asked him what they would do if there were a northern current instead. 'Then we will go anyway', he replied. Weather and ocean conditions are observed with the concern to conduct a safe fishing trip, but they do not hint precisely enough at the locality of fish.

The strongest incentive to go fishing on the first trip described above was the men's knowledge of the catches on the previous days and recent auction prices at the market. While on the day before the first trip, Ponnononam's catches were poor, they had been reasonably good on the day before the second. In both cases, the auction prices at the market had been satisfactory. On the second trip, inferences were made from the movements of trawlers and other *vaḷḷaṃ*-s. Abdu explains:

Abdu Puzhamkaraillathu: When three or five *vaḷḷaṃ*-s are seen moving and one *vaḷḷaṃ* is making a sudden turn than we can be sure that they saw some fish. If a boat is seen turning quickly it usually means that its men have seen some fish, then we also join them.

On other days, additional hints were relevant. For example, one day a bicycle trader brought the information about good catches and auction rates from the south. On another day, when catches had been poor on the previous days, the *vaḷḷaṃ*-s had not gone in the morning despite calm conditions. Then the *tarakan* received a telephone call announcing good catches in Nattika (ten kilometres north of Chamakkala). Three boats of the Ponnonam *vaḷḷaṃ* left within 15 minutes. As it was known in adjacent *kaṭavu̥*-s that Ponnonam is equipped with a mobile phone, other *vaḷḷaṃ*-s followed suit and also went to Nattika. For Ponnonam, this effort was in vain – it did not make any catch on that day.

Once at sea, a *vaḷḷaṃ* is part of a rapidly fluctuating and mutually responsive environment. This environment consists mostly of other *vaḷḷaṃ*–s and trawlers, but extends beyond the visible field of view, and may be affected, for example, by mobile phone calls from traders, the assumption of other teams having received such a call or even a weather forecast on the radio.

Making inferences about the knowledge of other *vaḷḷaṃ*-s is of particular interest, since these are aiming at the same species of fish. Observing the movement of trawlers is also of interest for artisanal fishermen, but since trawlers are thought to catch different fish species different conclusions are drawn from watching their movement.

Since trawlers are mainly fishing for prawns the movement of many trawlers in one particular direction is understood as a useful hint to go there as well. *Vaḷḷaṃ*-s may benefit from their speed. Observing such behaviour one cannot but feel that one is witnessing a race for a rapidly dwindling resource.

Lakṣyaṃ-s – the Signs of Fish

What does a shoal of fish worthy of being caught look like? In the descriptions of the two fishing trips a few of the signs[11] of fish (sing. *lakṣyaṃ*; sometimes called *aṭayāḷaṃ*; mark, symbol, signal) have been mentioned: the visibility of a characteristic pattern of ripples on the water surface and the lingering of seagulls at certain places in the sea. Sight is the most important sense for detecting shoals of fish. However, older fishermen claim that in the 'olden times' fish were caught after hearing or smelling their presence. One fish thought to be audible is *kōra* (jew fish), a predatory fish forming only loosely organised shoals. Small, shoal-forming fish are audible as well, as the second fishing trip described above has revealed. Talking about fishing at night, one man told me jokingly: 'oil sardine (*cāla*) is a friendly fish. It tells us when it comes'. The sound of a shoal of fish moving in the sea has been compared to the sound of wind. Some (older) fishermen insist that some fish can be smelt. For example, Kunhiraman Kochikkattu insists that *kannati* (an unidentified fish species) has a 'special smell' (*pratyēkata vāsana*) which stems from small bubbles (sing. *tari*) rising to the surface. When relating these matters to me, Kunhiraman, like other older men, emphasised that young men (and especially a foreigner like me) would not be able to make such subtle perceptions.

Common *lakṣyaṃ*-s are known by young and old men alike, but the skill of actually perceiving them is varied and depends on individual experience. In the daytime, shoals are detected by watching specific ripples in the surface of the sea or moving patches that appear shady or coloured. Descriptions of these signs vary. While most men agreed that a shoal of sardines is visible as a bluish or violet patch, shoals of mackerel were described to me as a mixture of green and blue or as brown. It is known from experience that these colours are not the proper colours of the fish. Rather, as Kunhiayyappan Kochikkattu explains, the visible colour is affected by the 'constitution (*prakṛti*) of the water through which the fish are seen'. Likewise, the changing illumination of the sea surface is known to affect the visibility of a shoal. The pattern of ripples caused by fish moving near the water surface is characteristic, but very difficult to see for untrained people. Another way to spot a shoal of fish is to watch for streaks of brownish water. These may be due to fish stirring up the mud (*mīn marikkunna*, lit. '[the] fish [is] turning [the mud]') while moving along the sea bottom. One may look for fish following such a streak to the end, where the dispersion is less and the trace fresh. Another hint at the presence of fish is the visibility of seagulls at the water surface, as was observed during the second fishing trip described above. Older men remember watching whales (*kaṭal āna*, lit. 'sea elephant') as sure hints for the presence of sardines in the sea – and possibly that of sharks, since both whales and sharks are said to feed on sardines.

Fishing at night is particularly common around new moon during the pre-monsoon hot season. Dark nights around new moon are needed to scan the sea for *kamaru̜*, a greenish light of varying intensity that is seen at the beach where the waves are breaking and where fish are present in the sea.[12] To the fishermen it seems that the fishes themselves glow in a green light, which some men ascribe to the salt on the fishes' scales.[13] The shape, movement and intensity of the glow may give a clue as to the kind of fish seen. Joseph explains that prawns (*cemmīn*) are recognised by wave-like movements in the light, which are apparently caused by the motion of the prawns' legs. A shoal of mackerel, he says, is visible by *kamaru̜* shining in a circular form.[14] The *kamaru̜* attributed to oil sardines is less specific, but usually regarded as fainter than that which is thought to be due to mackerel. Besides, there are variations in *kamaru̜* which are ascribed to season, specific currents or the saltiness of the seawater.

It should be added that fishing at night evokes an amazing otherliness of the marine environment. With the stars shining above in a dark sky and just a few lights of other boats seen on the horizon, the engine silently running at low speed, one is, now and then, gliding with the boat over patches of the diffuse greenish light of *kamaru̥* in the depths below. The sizes of shoals may be estimated by the light's luminosity. Once the decision to catch a shoal is made and the ring-seine net put around it, the fish begin noticing that their motion is now confined and they start dashing back and forth in the net. It looks as if a lightning storm occurs within the volume of water circumscribed by the net, a volume that is continuously reduced as the net is tightened by the fishermen pulling it onboard.

These visible signs are not only complex themselves. They are observed in situational contexts (season, time of day, heat, illumination, location and so on), which provide the frame to judge a specific observation. It is impossible to produce complete, all-encompassing verbal statements about the visible signs of fish in propositional form. The task of finding fish, and thus the answer to the question 'What does a shoal of fish worthy of being caught look like?' is context-dependent and pursued in a non-linguistic way.[15] It seems plausible that the fishermen do have an underlying idea (or 'mental model') of what constitutes 'a shoal of fish worthy of being caught'. The particular visible signs would then be interpreted according to the context, just like the Zafimaniry farmers of Madagascar studied by Maurice Bloch know 'what a good swidden is' without being willing or able to list all of its visible signs (Bloch 1991: 187). In fact, even more than in case of the swidden, the fishermen need to access their conceptualisation of the 'good shoal' quickly and process their sense perceptions rapidly in order to be able to act thereupon with success.

Not all the signs of fish may be useful for actual fishing. For example, many older and middle-aged fishermen claim that at night a shoal of fish may be found below a cluster of stars in the sky:

Joseph Parayattu: Stars (*nakṣatraṃ*) show the signs (*aṭayālaṃ*) of fish. It is said that where the stars cluster in an area [in the sky] and shine just like small spots then the fish should be beneath [under] them. On seeing [that] we should move to that side. But sometimes the stars will be further away than we usually go. Then we should move in our area in search of fish. Most often we do not go beyond our area but sometimes we may try. We cannot measure the angle (elevation) of the stars and the distance may be too great. Then we will move around at the limit of our area. Sometimes we will get some catch and sometimes not.

Similarly, Kunhiayyappan Kochikkattu insists that, formerly, one would surely catch fish by following this rule (*niyamaṃ*), but he laments that old rules such like this do not seem to work any more in the present advanced times (*purōgamiccu̥ kālaṃ*).

What kind of relationship links stars and fish? Asko Parpola (1994), who claims that the language of the Indus civilisation may be related to Dravidian languages of South India, provides us with a hint. He notices that in the vocabulary of many of these languages, including Malayalam, the words for 'fish' and 'star' are homophonous: *mīn* (ibid.: 182; see also Gundert 1995 [1872]: 822). Is the *lakṣyaṃ* misconceived because of this linguistic slippage? Or does it hint at a conceptualisation linking sea and sky?

With other clues lacking, the correspondence between clusters of stars in the sky and shoals of fish in the water below may remind one of Italo Calvino's *Invisible Cities* (1997). In this novel Marco Polo reports to Kublai Khan each time he returns from his travels, portraying the cities he has visited. Upon one of these returns, the record is made of a city called Andria: '[This city] was built so artfully that every street follows a planet's orbit, and the buildings and the places of community life repeat the order of the constellations and the positions of the most luminous stars: Antares, Alpheratz, Capricorn, the Cepheids ... "Our city and the sky correspond so perfectly", [explain its citizens], "that any change in Andria involves some novelty among the stars". The astronomers, after each change takes place in Andria, peer into their telescopes and report a nova's explosion, or a remote point in the firmament's change of color from orange to yellow, the expansion of a nebula, the bending of a spiral in the Milky Way. Each change implies a sequence of other changes, in Andria as among the stars: the city and the sky never remain the same' (Calvino 1997: 150-151).

The Art of the Skipper

The above descriptions illustrate the central role of the *aryakkāran* (skipper) in the fishing trip. It is in his person that cognitive and social skills meet. Is the *aryakkāran* the most skilled fisherman on board? If so, in which particular way is he skilled?

When I asked fishermen in Chamakkala the question 'What makes a good *aryakkāran*?' their answers were stereotypical. An *aryakkāran* needs to have the innate desire for fishing, make a sound judgement about the day's weather conditions and be courageous to make a bold decision, needs sharp vision to spot fish in the sea, he needs to be able to conduct the fishing trip, gain the trust of the crew, retain this trust when there may be no catch for days and make the team members be punctual and reliable in their conduct. This is a broad mixture of cognitive and social skills. In many cases a unit's conduct is said to be its *aryakkāran*'s conduct writ large. Then, how does one become the *aryakkāran*? This is Abdu's answer to this question:

Abdu Puzhamkaraillathu: Like going to school there is a training period. At first we do not have the ability to find and catch the fish. The old ones may be doing that work and we will sit in some unimportant corner [of the boat] and watch how the others are doing the work. In this way we are studying the things. After the elementary studies we are sent to the upper primary school. Like that in a *vañci* we should be promoted to the next post. Thus we will become the *aryakkāran* in the end.

In this description Abdu blends the olden days, when he himself began as a fisherman (thus the reference to the *vañci*-s), with the present time (the term '*aryakkāran*' was introduced with the *vaḷḷaṃ*-s).[16] Learning to become the *aryakkāran* thus means to achieve a particularly high level of skill which is, in its essence, not different from every fisherman's skill. Abdu emphasises the importance of gaining experience through the participation in fishing with others. This acknowledges the gist of practice theories of learning, popular in psychology since the 1980s, which emphasise the situatedness of learning (Lave and Wenger 1991; see also Pálsson 1994). Indeed, fishermen in Chamakkala (and probably at most other places) do seem to learn their practical skills through 'legitimate peripheral participation', as described by Lave and Wenger (1991: 35–37): watching what the skilled people do and increasingly participating oneself. In this way, education is not a one-way process, but rather a mutual experience. Skill is accumulated through experience. If so, part of the *aryakkāran*'s skill is akin to that of an expert in chess, who is 'not necessarily somebody with supreme logico-mathematical skills, but rather someone who has stored a comprehensive and organised store of total or partial chessboard configurations' (Bloch 1991: 188). In Chamakkala, in many situations of deciding whether to embark on a fishing trip or which decision to make at sea, previous experiences are referred to. These are often those which are shared by the crew members present. Abdu's decision to go fishing upon watching the Sudarsanam *vaḷḷaṃ* doing so is such a reference. Judging the general situation of fishing on a particular day and considering any particular shoal of fish are others. To have acquired such a stock of knowledge and experience is part of the *aryakkāran*'s skill. Like Abdu, Joseph stresses the importance of experience as the major factor in developing the cleverness (*yukti*) needed by an *aryakkāran*. 'The cleverness of an *aryakkāran*', says Joseph, 'is not to let the fish escape. Fish don't want to be caught, they will try to escape'. Fishing therefore means to 'outwit' the fish, a task needing varied skills, since some fish are smart (like oil sardines), while others are stupid (like *cōrakanni*), as Joseph explains.

No single trip can be predicted in detail and thus the ability to improvise is paramount to the success of a team and the skill of an *aryakkāran*. Both of the above trips have shown how much flexibility is needed (for example, to adapt to being surrounded by different patterns of trawlers). Taking this into account, one may call any fishing trip a unique performance (rather than a rehearsal of a predetermined, routinised pattern), the *aryakkāran* being the central performer, but not the only one on board. As seen above, technical and cognitive skills and tasks are distributed among the team members, but it is the *aryakkāran*'s task to coordinate them (see Hutchins 1995).

Many of his skills can be acquired by all fishermen, as Manoharan Eranpurakkal and Vijayan Kovil Tekku Valappil reminded me, drawing attention to what distinguishes the *aryakkāran* from other crew members:

Manoharan Eranpurakkal and Vijayan Kovil Tekku Valappil: They are the *aryakkāran*-s, but we have also done the work on the boat for so many years. We also know what is the time to place the net. We know the things which an *aryakkāran* knows, but they are brave enough to stand at the *aniyaṃ* (front) and control the work. Being brave enough to stand there every unskilled worker can become an *aryakkāran* within five months.

There is general agreement that Joseph has above-average cognitive skills and is very knowledgeable. And he is able to gather the team behind him. Even when catches have been small for a prolonged period, he himself is usually not blamed. Joseph is well liked in the village. He often jokes and is generally counted as a cheerful person. As the *aryakkāran* of Chamakkala's principal *vaḷḷaṃ* unit, he is one of the most authoritative men on the beach. But Joseph is a non-local Christian in a team of local Hindus (and two Muslims). He deals with this awkward situation by taking charge of the duties expected of a Hindu *aryakkāran*. This demands of him being present at rituals in the temple of the unit's *mantravādi*. In addition, his horoscope is regarded as fateful for the unit's conduct as a whole, and is investigated as such.

Summing up, it seems that cognitive skills and knowledge of the fishing process are known more or less by all fishermen. But their work must be coordinated. Symbolically it is the braveness of standing in the front, the willingness to accept the role of social leadership and stand by it with confidence, which distinguishes the *aryakkāran* from the other men on board.

Notes

1. See Hornell (1937: 58–61), Mathur (1978: 162–167), Kurien and Willmann (1982: 5–6). For an illustration see Hornell (1937: 59). According to SIFFS (1992: 23), previously the *kolli vala* 'has been the main gear for nearly half of Kerala's fishermen'.
2. For instance, I was told by a man that he worked on his father's *vañci* in his youth. For years, the crew consisted of himself, his elder brother, his father, his father's younger brother and a Muslim neighbour.
3. Rajan (1993: 34) notes that only sixteen men were required for fishing with the *kolli vala*. I suspect that this number is the sum of the crews from two larger *vañci*-s of eight men each. Rajan probably did not count the men on the *vañci*-s who were generating the noise.
4. Either these names were descriptions of the set (such as the *ancụvalakkar set* which was composed of five – Mal. *ancụ* – two-boat units), or they could relate to stories relating its members. Thus, one set was called *kākkana vētikaḷ*, lit. 'crow killers', because of a story relating to a member hitting a seagull (*poṅṅaṭṭi*, not a crow, *kākka*) in flight with an oar. The seagull had attempted to take fish from the net. The third set was called *lōpụ set* (after *lōpụ vala*, gill net).
5. Onam is the most elaborate annual festival in Kerala (see Tarabout 1986: 78–82; Osella and Osella 1996). Marking the end of the south-west monsoon it commemorates the annual return of Mahabali, once the ruler of Kerala in a period of prosperity.
6. At this time they were equipped with *taṅṅụ vala*-s, which lacked rings at the bottom. The *ring vala* was introduced in 1986 or 1987 in this part of the Kerala coast.
7. While net sizes had been previously measured in *marụ* (denoting their length and depths), they are now referred to in terms of their weight (in kilograms) more and more often. At the Matsyafed-run Vyasa stores, where most local fishermen buy their nets, prices are commonly determined according to the weight of nets. Large ring-seine nets prove to be cumbersome gear, particularly in Thrissur District. Being too heavy to be carried by hand, they are often brought to the boats on trolleys. In addition one may note that the global unit 'kilogram' appears to replace the local unit '*marụ*' (see Appadurai 1989; Scott 1998: 25ff.).
8. Lack of success in fishing may quickly develop into a debt trap, since units with (perceived) little success are forced to accept loans with higher interest rates. This may lead to the failure of a team.
9. Even though anybody may request to be accepted as a worker on a *vaḷḷaṃ*, the implicit consensus of the shareholders is required. More or less subtle mechanisms are in use to exclude somebody from participating whom the shareholders do not want to do so. The *aryakkāran* has the right to exclude somebody from participating, but, if this does not reflect a more general opinion of the crew members, the *aryakkāran* may weaken his own position. Such discussions occasionally cause significant tensions in all *vaḷḷaṃ* units.
10. Songs with the refrain '*...ailassa*' are common work songs among the fishermen. The text of one of these songs has been recorded by Sunish (1995: 106) as: 'Elelam, let it lift up / Pull it together, oh companions / Now it will rise up / Elelam, pull together/ Pulled together, the mountain will rise as well' (translated from the Malayalam by L.S. Rajagopalan, Thrissur).
11. By sign I understand the perceptible (visible, auditory and olfactory) clues to the presence of fish in the sea. Semiotically, these are indexical signs as they are linked to the object of signification (fish) by contiguity (see Daniel 1984: 31).
12. Marine biologists have identified the *kamarụ* as bioluminescence produced by the bacterium *Vibrio* sp., which is commonly found in the Indian shore waters of the Arabian Sea. The glow is enhanced where *Vibrio* is exposed to oxygen, as is the case where the waves are breaking and when fish move through the water. I thank Dr V. Chandrika (Central Marine Fisheries Research Institute, Kochi) for explaining the biology of Vibrio to me (December 1999).

13. Some men told me that this salt absorbs the sunlight during the daytime and then re-emits it at night.
14. According to Dr Chandrika, this circular shape is due to the fact that the content of oxygen dissolved in the water is smaller in the centre of a shoal, where it has been used by the fish.
15. If one asks whether a particular shoal of fish is worth being caught, even the temporal market price of fish affects the *aryakkāran*'s judgement of a shoal.
16. On a *vañci*, instructions were given by the *sranku̧*.

4 Sea and Fishes

How do Chamakkala fishermen conceive the marine environment? In which terms do they communicate about it? In this chapter I present my ethnographic material in terms of local concepts and supplement it by commentaries and additional observations. I am interested in consensus and dissent, as well as in the men's styles of reasoning.

'The Sea is a Woman, the Mountains are a Man'

Chamakkala fishermen commonly refer to the sea as female and as mother. This is what Krishnankutti Kochikkattu did in the conversation at the beginning of Chapter 1. He and others refer to the sea as *amma* (mother) or *kaṭalamma* (lit. 'sea mother'): '*Kaṭalamma* is lying down spread out like this' (*kaṭalamma iṅṅanu̥ vistarāmayi kiṭakkunnu̥*). Taken literally this would mean that the sea is identical with *kaṭalamma*. Identifying *kaṭalamma* with the sea is often emphasised by older fishermen, who note that when going on a fishing trip small prayers had been addressed to the sea as *amma* ('mother'):

Kunhiraman Kochikkattu: First we carry the boat to the sea. Then the boat is ready to start. After that we touch the seawater and taste it and after that we move to the sea. This is *kaṭalamma*, yes, *kaṭalamma*. It is mother (*amma*). Some water from mother is taken and drunk and then we move to the sea. Some practised this and some did not. Yet it appears to me that the good and honest people always used to do so.

This custom reminds one of the performance of ritual in a Hindu temple, where, after the conduct of *pūja*, worshippers take some of the food that has been offered to the deities as *prasadaṃ*. This *prasadaṃ* is believed to contain some of the deity's *śakti* (power), which is incorporated upon ingestion.

Referring to the sea as *kaṭalamma* is common when its protecting and providing (i.e. 'motherly') character is stressed ('It is *amma*, therefore ...'). Is *kaṭalamma*, then, a goddess? Such a notion would not be unexpected, since throughout India the land is similarly associated with Bhudevi, the Earth goddess (Fuller 1992: 44; Osella 1993: 455). I have not come across any temple or shrine for Bhudevi and, likewise, there is no temple or shrine consecrated to *kaṭalamma* as a deity. The emphasis put on offerings at several local temples for Bhadrakali and Bhagavathi is explained by these deities' close relationship with the sea and their particular concern with the well-being of the fishermen. Some older men emphasise that *kaṭalamma* as a deity is identical with the sea. As Kutti Kovil Tekku Valappil, a retired fisherman in his seventies, says:

Kutti Kovil Tekku Valappil: See: our mother is giving birth to us. That is a truth of life (*jīvitattinṟe satyaṃ*). In the same manner *kaṭalamma* is important to us. Do you know who *kaṭalamma* is? She is a goddess (*daivaṃ*), the one lying down in the west (*ī paṭiññātụ ī kiṭakkunatụ*).

Perceiving the sea as female is common among fishermen throughout India (see Mathur 1978; Ram 1991; Bavinck 2001), while in (north Indian) Brahmanical discourse and in classical Sanskrit poetry the sea is generally described as male, the rivers being his wives (e.g. Nandargikar 1971: 399–400; Feldhaus 1995: 54). Grammatically, the Sanskrit word for 'sea' or 'ocean' (*samudra*) is male, while Sanskrit names of rivers are always female (e.g. *ganga*, *yamuna* or *narmada*). As a Dravidian language, Malayalam lacks gender forms, and thus a local tradition in which the sea is female (and rivers are male) is reconcilable with the grammar of the local language.

In the conversation at the beginning of Chapter 1, Krishnankutti had left the gender of the mountain water (*mala veḷḷaṃ*) open. Some fishermen in Chamakkala regard it as being male. If this view is adopted, the myth told by Krishnankutti turns into a conversation between two waters, those of the sea and those of the mountains.[1] All fishermen in Chamakkala agree about the sea's female gender. But whether the mountains are considered as male, as Krishnankutti asserted, is at the most suspected but more often questioned or even denied. The association between the Himalayas and the Western Ghats (the Malayas), which is made in a popular Ayurvedic compendium, the *Ashtangaridhaya Samhita* (Zimmermann 1987: 70) supports this claim. Mount Meru in the Himalayas is considered as Lord Shiva's preferred abode for meditation. The two most popular destinations of pilgrimage visited by Hindu fishermen from Chamakkala are the shrine of Ayyappan in Sabarimala (south Kerala) and the one of Murukan/Subramoniam in Palani (Tamil Nadu). Both deities are sons of Shiva and their temples are situated in the mountains.[2] In Kerala, most of the important temples for the Goddess are situated near the sea. Two examples from the vicinity of Chamakkala are the Bhadrakali temple at

Kodungallur and the Bhagavathi temple at Chottanikkara (Vaidyanathan 1982). Feldhaus (1995: 26ff.) notes that, in Maharashtra, Shiva lingas (symbolic images resembling a phallus) are commonly seen on mountain summits in the vicinity of the sources of rivers. In the sense indicated by Eck (1998: 167) this seems to emphasise the association between these rivers and the (mythical) Ganga. Similarly, other mountains are identified, explicitly or implicitly, with Shiva, a Shiva linga or a Shaiva deity (Feldhaus 1995: 26 and 37: footnote 21).

Conceiving the sea as *kaṭalamma* has become known in Kerala beyond the fishing community through Thakazhi Shivasankaran Pillai's novel *Cemmīn* ('Prawn'; 1962 [1956]). A popular movie picture of the story was made in 1965. In his novel, Pillai (who was a high-status caste Nayar) refers to *kaṭalamma* as the 'goddess of the sea' (1962: 81, 209) and portrays her as a moral authority to which the members of the fishing community have to submit, *kaṭalamma* demanding supreme moral conduct. The misconduct of a woman during her husband's absence in fishing may result in the husband's injury or death at sea. In the novel's tragic end, Palani has gone fishing, while his wife, the heroine Karuthamma, is visited by Parikutti, a Muslim trader for whom she has had strong feelings since her pre-marriage days. The next day, Karuthamma and Parikutti are found dead on the shore. Palani has been lost in a storm: 'The palace of the goddess of the sea was at the bottom of the deep sea. There the sea goddess was enshrined. He [Palani] had to get there through a whirlpool, a whirlpool which made the whole sea churn round in circles, knocking at the gates of the sea goddess' abode' (Pillai 1962: 209).

When Pillai died in April 1999, obituaries praised him as a social realist writer who had always taken much effort in researching his characters' background carefully.[3] Since Pillai lived in a village near the shore (in south Kerala's Allapuzha District), one may surmise that the notion of *kaṭalamma* as described in the novel is inspired by the concepts of local fishermen.[4] Shortly after the publication of the novel, Pillai was attacked by V. V. Velukutti Arayan, a well-known political (communist) activist of the day, in a polemical pamphlet (Velukutti Arayan 1956). Pillai's notion of *kaṭalamma* was at the centre of the critique. Velukutti Arayan felt his caste denigrated by Pillai's claim of their alleged worship of *kaṭalamma* as a non-Sanskritic deity. He demanded that one should not take the concept of *kaṭalamma* literally (as Pillai had done) and considered the alleged moral authority of *kaṭalamma* as oppressive against women's liberty. Velukutti Arayan did agree, though, that (in south Kerala) in 'sea songs and other songs of valour the sea is "personified" (*sarupatvaṃ*) as sea mother, queen etc. just as a poet addresses trees and creepers' (Velukutti Arayan 1956: 26). Thus, what in Pillai's novel is a living force of moral authority is considered by Velukutti Arayan as mere poetic speech.

I am not sure whether Pillai ever responded to Velukutti Arayan's polemic. The difference between their viewpoints touches on ethnographic and linguistic problems which will be central to this chapter and the following ones. Does talk of the 'sea mother' simply mean to use a metaphor? Or do the fishermen consider the sea as mother in another sense? If so, on which grounds may the sea be considered as mother? Where, and how, can we locate the border between literal and figurative speech? Or, more generally: What must the fishermen believe about themselves and their surroundings for their metaphors to work?

At first sight, Velukutti Arayan's reference to poetic language may remind one of Giambattista Vico's (1668-1744) understanding of figurative language. This Italian philosopher regarded the attribution of human bodily features, senses and passions to inanimate objects as the 'language of the first poets' and regarded it as an early effort to make sense of an otherwise incomprehensible world. In *The New Science*, Vico writes:

> It is noteworthy that in all languages the greater part of the expressions relating to inanimate things are formed by metaphors from the human body and its parts and from the human senses and passions. Thus, head for top or beginning; the brow and shoulders of a mountain, the eyes of needles and potatoes; mouth for any opening; the lip of a cup or pitcher; the teeth of a rake, a saw, a comb; ... the flesh of fruits; a vein of rock or mineral; the blood of grapes for wine; the bowels of the earth. Heaven or the sea smiles; the wind whistles; the waves murmur; a body groans under a great weight. The farmers of Laetium used to say the fields were thirsty, bore fruit, were swollen with grain; and our rustics speak of plants making love, vines gone mad, resinous trees weeping. (Vico 1968 [1744]: 129)

Note that Vico implicitly assumes inanimate things and (animate) human beings to be clearly distinct. They seem to make up two categories which, once established, can be related to each other. Vico proposes two ways of doing so. In a first group of examples (from the 'head' to the 'bowels of the earth') he attributes 'properties of the human body' to inanimate entities. These examples could be termed catachreses, defined by Black (1962: 33) as 'the use of a word in some new sense in order to remedy a gap in the vocabulary'. One may also call them anthropomorphisms, since these expressions describe the shape of non-human things in terms of the human body. The examples of the second group (from 'the smiling heavens' to the 'weeping trees') may be called proper metaphors, since they do not just fill a lexical gap, but rather help in 'understanding and experiencing one kind of thing in terms of another' (Lakoff and Johnson 1980: 5).

Vico was well aware that metaphors may serve as means to make the surrounding world comprehensible, regardless whether they are used consciously or not. He continues:

> All ... is a consequences of our axiom that man in his ignorance makes himself the rule of the universe, for in the examples cited he has made of himself an entire world. So that as rational metaphysics teaches that man becomes all things by understanding them, this imaginative metaphysics shows that man becomes all things by not understanding them; and perhaps the latter proposition is truer than the former, for when man understands he extends his mind and takes in the things, but when he does not understand he makes the things out of himself and becomes them by transforming himself into them. (Vico 1968 [1744]: 129–130)

Thus, figurative language may be used in a merely poetic fashion, but it may also be fundamental to understanding, or, when used in everyday speech, be essential to communication. Keeping these possibilities in mind, and paying attention to the border between the figurative and the literal, I shall now turn to the local conceptualisation of the sea's spatial structure.

The Sea is Lying Down

In everyday conversations, the sea is usually referred to as the *paṭiññāṟụ* (west). The Malayalam word *kaṭal*, which literally means 'sea' or 'ocean', is used less often. As previously mentioned, fishing is usually referred to as *paṭiññāttụ pani*, the '(manual) work of the west'.[5] Referring to the sea as the *paṭiññāṟụ* is an essentialist use of a word which in literary language plainly denotes the cardinal point: *paṭiññāṟụ* (lit. 'west', idiom. 'sea') is opposed to *kiḻakkụ* (lit. 'east', idiom. 'land'). While travelling in the sea, though, *paṭiññāṟụ* and *kiḻakkụ* are used in their common, relational meaning as cardinal points.

Even though the term *paṭiññāṟụ* can be synonymous with 'sea', it is subtly distinguished from the entity 'sea, ocean' (*kaṭal*). Thus, when I enquired about the structure of the sea, answers often included statements like *kaṭal paṭiññāṟụ* side-*il kiṭakkunụ* ('the sea is lying in the western side')[6] or *kaṭal iṅṅanụ kiṭakkunụ* ('the sea is lying down like this'). The verb *kiṭakkuka* (to lie [down], rest, dwell; Gundert 1995: 247; Warrier et al. 1999: 267) expresses the passive permanence of the sea as being localised in the west. It contrasts with the verb *nilkkuka* (to stand, remain; Warrier et al. 1999: 592), which describes the activity of standing, usually for a limited time in a certain place or locality. It is commonly used when talking about the places (*stalaṅṅaḷ*; sing. *stalaṃ*) of animate beings (like humans and fishes) and inanimate beings (like fishing boats) in between movements. We shall see in the next section that in certain contexts *nilkkuka* may be used to refer to the situatedness of the sea.

As is typical for central and northern Kerala, in Chamakkala the sea bottom is sandy and descends towards the west with a shallow slope (Kurup 1972: 6; PCO/SIFFS 1991b: 6). This slope is considered to be more or less constant up to the westernmost places reachable with the boats. Thus, to know the depth of the water at a given locality gives an idea about this place's distance from the beach. The depth (*tārca*) at a given place is expressed in *bhagaṃ*. One *bhagaṃ* is the distance between stretched-out hands or about a man's height (approximately 1.7 metres).[7] It could be measured with a stone attached to a rope which could function like a sounding lead. More often it would be guessed from experience, especially by watching the coast or some mountains further in the east. Because of the assumed constant slope of the sea bottom, the *bhagaṃ* scale has been used to denote the distance of boats in the sea.

In calm weather conditions, when little or no surf is present, the *vaḷḷaṃ*-s may be anchored in the *arāyi* and small *vañci*-s would be fishing there with gill nets. Some men refer to this surf zone as the *paṭi* (lit. 'door, gate, threshold'). Together, *kaṭappuṟaṃ* and *arāyi/paṭi* constitute the shore (*kara*). The latter is an ambivalent term: fish may be 'standing' in the *kara*, the *cākara*, a coastal mudbank that forms seasonally during the south-west monsoon, may 'dwell' in the *kara* and even the boats on the beach are in the *kara*, which thus marks the transition between land and sea. Some people refer to the sea immediately west of the *arāyi/paṭi*, at depths of three to about five *bhagaṃ*, as the *iṭa maddyaṃ* (lit. 'space in between'). Others deny that there is any particular name for this part of the sea. Still further away to the west is the *puṟaṃ kaṭal* (outer sea), sometimes referred to as *mānaṃ kuṭiya kaṭal* ('sea of big volume'), *mēle kaṭal* ('upper sea') or *āḻamuḷḷa kaṭal* ('depth-having sea'). It reaches as far to the west as is imaginable, with ever-increasing depth. There is no convention as to where the *puṟaṃ kaṭal* begins. According to older fishermen it begins at a depth of about five, six or eight *bhagaṃ*. Some middle-aged fishermen disagree and concede that what is properly called the *puṟaṃ kaṭal* has been considered to be further in the west since the advent of motorised boats. Recently, the unit *bhagaṃ* has been supplemented by expressing this distance in kilometres. The conversion is simple: since the sea is supposed to drop off towards the west by one *bhagaṃ* per kilometre, the numerical value for a specific place remains the same but the unit is changed from *bhagaṃ* to kilometre.

For older and younger fishermen the *puṟaṃ kaṭal* is that part of the sea which is rarely visited, due to difficulty of access and lack of economic benefit. Older fishermen commonly emphasise that bravery (*dhairyaṃ*) had been required to reach the *puṟattu* (lit. 'outer side', i.e. the *puṟaṃ kaṭal*). Changing weather conditions and currents could make a fishing trip to the *puṟaṃ kaṭal* dangerous and uncertain. The hardship of rowing (inf. *taṇṭu valikkuka*, lit. 'pulling the oar') over this distance was an additional burden, even though sometimes sails had been used. In contrast, middle-aged and young men working on motorised boats emphasise that fishing trips to the (redefined) *puṟaṃ kaṭal* are economically not viable, due to travelling times and expenses for fuel.

The sea bottom (*aṭi*), while generally thought to be sandy and muddy, is not completely devoid of structure: the places of several reefs and shipwrecks are known. While it is thought that some fish gather at these places, knowing them is less important for fishing itself but rather in order to avoid damage to the nets. Mountains of sand (*māṭu*) and lakes or lagoons (*tāṭakaṅṅaḷ*, sing. *tāṭakaṃ*) are further elements of the underwater topography. Most of the lakes or lagoons are situated far away in the *puṟaṃ kaṭal*, inaccessible to *vañci*-s and *vaḷḷaṃ*-s. They are thought to be depressions in the sea bottom in between reefs, believed to be rich in fish, particularly prawns (*cemmīn*). While known only through hearsay from fishermen working on trawling boats, the properties of the lakes are claimed to match experience and observation from the immediate coastal environment. Vasu (late 60s, recently retired from working on the *vaḷḷaṃ*-s) explains:

Vasu Kochikkattu: When we move to an *aṟappa* [a small river inlet] we can see deep and shallow places. To fill them, either sand (*mannu*) or mud (*cēṟu*) will move there. If they are filled with mud it will be a lake (*tāṭakaṃ*) itself. From there you can get a lot of prawns (*cemmīn*).

As mentioned above, both the depth in the sea and the distance from the beach are measured in *bhagaṃ*-s. The vertical structure of the sea may also be distinguished in a more vivid way into the surface of the sea (*kaṭalinṟe mukal*), the upper part of the sea where fish can be seen from above (*kaṭalil*, lit. 'in the sea'), the invisible region immediately below (*tāḻe kaṭalil*, lit. 'under the sea'), the 'even more invisible' region below that (*nīre tāḻe kaṭalil*, lit. 'beneath the sea') and, eventually, the bottom of the sea (*kaṭalinṟe aṭittaṭil*). These terms are used in making reference to fishing, e.g. in descriptions of where fish are seen or where their presence is surmised.

The Sea is Standing Up

In our first conversation, Krishnankutti Korupattu complained about the impact of trawling boats and foreign fishing vessels. Because of their presence in the *puṟaṃ kaṭal*, he pondered that many fishes would not come to the coast any more:

Krishnankutti Korupattu: Nowadays the ships and boats move even to the *puraṃ kaṭal* for fishing. Because of this the fishes are rising [and] rising [and are] standing [there] (*a enna kattinụ mīn kayaṟi kayaṟi nilkkukayāṇụ*).

Previously, when talking about the fishes' behaviour, Kunhiayyappan Kochikkattu (eighty years old, retired from fishing about twenty years ago) had remarked:

Kunhiayyappan Kochikkattu: From the western sea fishes are coming [by] descending (*atụ paṭiññāṟan kaṭalil ninnụ mīn ayiṭṭē iṟaṅṅi varikayullụ*).

The verbs *kayaṟuka* ('to enter, rise, increase, mount'; Warrier et al. 1999: 232) and *iṟaṅṅuka* ('to descend, alight, ebb, become less'; Warrier et al. 1999: 147) are commonly used in referring to the eastward and westward motion of fish in the sea, respectively. The same verbs are used in an everyday context when describing motion into or out of a confined, and usually elevated, space, e.g. a room (*muṟi*).[8] Examples are: *muṟiyil kayaṟuka* ('to enter a room') and *muṟiyil ninnụ iṟaṅṅuka* ('to leave a room').

Is the use of these verbs solely a matter of convention, as these examples suggest, or do they literally imply, in the fishermen's imagination, that the sea rises toward the west? Kumaran Eranpurakkal, in his early seventies, gives this answer:

Götz Hoeppe: The fish will go up and up. What does it mean? (coll. *mīn kayaṟi kayaṟi pōkkuṃ eṇāl eṅṅāṇụ?*)
Kumaran Eranpurakkal: That they move away from the coast, that is all (*atāyatụ karayil ninnụ pōkkuṇụ eṇụ mātraṃ*).
Götz: Does the sea go up towards the west? (*kaṭalụ paṭiññāṭṭụ pōkkuṃ tōṟuṃ poṅṅiyāṇụ varunatụ?*)
Kumaran: That is our thought (*uddēśaṃ*). Of course, you may ask: how can water move up towards one end? But if we move five *bhagaṃ* [into the sea] then we can see the mountains in the east, except for the small ones. When we reach ten *bhagaṃ* then we can see all the mountains.

Like all the older, many of the middle-aged and some of the younger people I talked with, Kumaran asserts the sea to be rising (*pongal*) above land level towards the west. Even though he agrees that this claim runs counter to everyday experience, he feels it to be vindicated by the empirical observation of the mountains being visible from the sea. Others support this claim by arguing that it would always be more difficult to row a certain distance towards the west than to return towards the east, thereby utilising the analogy of the physical effort needed to climb a mountain and the similar effect to mount (inf. *kayaṟuka*) the sea.

Even though the rise of the sea is considered by many to be proved by empirical observation, it is nevertheless considered a remarkable fact. Vasu Eranpurakkal, who, despite his age of sixty-eight, is still working on a *vaḷḷaṃ* and in his own three-man *vañci*, explains:

Vasu Eranpurakkal: The sea is standing (*nilkkuttatụ*) above us. This is a great wonder (*atiśayaṃ*). The sea is higher up than we are! Water at this height can flow towards us, shouldn't it? If you pour water into a vessel then it will rest in its bottom. But the sea is up high and we are below that in the plain land. Because of this the seawater can flow here. This [the standing of the sea] is by the might of God (*daivattinṟe kaḻivụ*). We are saying that this is due to the might of God.

Note that Vasu refers to the sea as actively standing (inf. *nilkkuka*) in contrast to the passive lying down (inf. *kiṭakkuka*). Like Kumaran, Vasu contends that the rising of the sea contradicts common sense, since water usually flows down. In attributing this rise to the might of God[9] (*daivattinṟe kaḻivụ*) he asserts that God consciously protects the people who live in the coastal strip. This resembles the myth of the dialogue between the sea and the mountain water (Chapter 1). In either case people in the coastal area are consciously protected, either by the sea or by God (*daivaṃ*). This world is inherently unstable, since the continuous blessing is required. The sea's active and conscious protection of humans is often traced to its *satyaṃ* ('truthfulness', see Chapter 5). Moral misconduct may obscure the blessing of God or the sea's *satyaṃ* and may result in a flood. Several older fishermen assured me that the flood which had occurred recently in Orissa (October 1999) must have been due to the improper moral conduct of the people living there.

Some people referred to the *vikṛti* of the sea (*kaṭalinṟe vikṛti*) in accounting for its rising. The word *vikṛti* may be translated as 'change, abnormality, vicious attitude or mischief' (Warrier et al. 1999: 972) and is an attribute commonly used in contexts of affectionate familiarity between persons. Thus, a mother may tell her annoying children: *vikṛti kānikkātirikkụ!* ('stop making nonsense!'). In this light, *vikṛti* is considered as a playful, but never malevolent, disturbance. If the sea's rising is called a *vikṛti*, then it is a highly benevolent one for the fishermen.

Lately, the view of the sea rising towards the west has been disputed (see Chapter 7).

The Lotus Sea (*tāmaru̩ kaṭal*)

On one day in late February, at the beginning of the hot season (*vēnal pakal*), Vasu Kochikkattu complained that hardly any catch had been made recently. None of the boats had gone fishing that day. The heat was beginning to be uncomfortable. It would keep increasing daily. Vasu, Suresh and I were sitting on the beach, in the shade of a coconut palm. Here at least a fresh breeze could be felt. Some fishermen were asleep, others were playing cards or were talking. The nets had been mended beforehand. It was an idle day. When I asked Vasu whether he would expect more fish to come in the south-west monsoon, he gave an elaborate answer:

Götz Hoeppe: In the monsoon, will there be more fish in the sea?

Vasu Kochikkattu: At the time of the monsoon the fish are multiplying (inf. *perukkuka*), they are having the small ones. For *paṟṟal* ('spawning'; lit. 'sticking, adhering') the fish move to the upper sea (*mēle kaṭal*) by the end of *kuṃbhaṃ* month [February-March]. That is the shelter for them. Our rule (*niyamaṃ*) says that they are moving to our Lotus Sea (*tāmaru̩ kaṭal*) in the west. Beneath the Lotus Sea the fishes are laying their eggs. The eggs are multiplying (inf. *perukkuka*) there. From there by the movements of the water [currents] and by playing the fish are coming to the coast. By that time we can meet [i.e. see] these fish.

Götz: What is the Lotus Sea (*tāmaru̩ kaṭal*)?

Vasu: The Lotus Sea! There the sun's rays do not reach the water, just as in a big pond full of lotus it [the water] should always be very cold, because in this pond the sun's rays only touch the green leaves of the lotus and do not reach the water. The sunlight is touching these leaves. There must be a country (*rajyaṃ*) or place (*stalaṃ*) where they [the fish] are safe to spawn. In that place they are hiding happily (*ceniṭṭu̩ sukhamāyiṭṭu̩*). Beneath the leaves they are not disturbed by sunlight, wind or anything like that. ... It should be as if we are lying in a safe house. No sunlight, no sun heat, no problem at all! There it should be cold at any time. If I am describing this in our words: A fish which comes from the *puṟaṃ kaṭal* is used to prepare a curry. It will have a certain taste (*svādhu̩*). But if the fish is from the Lotus Sea it should have the smell of lotus (*tāmarayuṭe vāsana*). You will not be able to notice this, but we can easily identify it. If we taste that curry we notice the smell of lotus (*tāmaru̩*). For example: if some kerosene (*manana*) is dropping onto the fish in the boat then that fish should have the smell of kerosene. By this [i.e. by the smell] it is clear that the fish is living in the Lotus Sea and has just come from there. If we catch the fish of the Lotus Sea and

> make a curry from it then we can understand this property [the smell of lotus]. Thus we are finding that this fish is not one of those which move around here, but that it has come from the outer sea (*puṟaṃ kaṭal*).

Vasu emphasises the parallel between the fish and the fishermen in their preference for cool, refreshing places. That he mentions the Lotus Sea (*tāmaru kaṭal*) in this conversation at the beginning of the hot season may not have been by chance but seems to be related to the particular situation, in particular his own discomfort with the rising heat. Descriptions of the Lotus Sea are fairly uniform and similar to Vasu's: it is said to be a distant part of the sea, far in the west, which is covered with lotus petals in the hot season (*vēnal pakal*). No sunrays can traverse this cover. When the heat of the seawater rises at the beginning of the hot season, the shade below the lotus flowers is the only cool region in the sea. This makes it an attractive destination for the fish. Several fish species (especially *ayila*, 'mackerel') are supposed to spawn in the Lotus Sea. They form groups (*kūṭṭaṃ* or *pelappu*), in which they descend towards the coast at the beginning of the south-west monsoon (*kālavarśaṃ*). At this time, the rough sea may cause the stems of the lotus flowers to break and the Lotus Sea to dissolve until the end of the monsoon.

When I had enquired about the structure of the sea previously, all fishermen had called the *puṟaṃ kaṭal* its furthest end, but none had talked about the Lotus Sea in this context. Accounts of the Lotus Sea always implied the fishes' desire to go there in the hot season. Their taste (*svadhu*) and smell (*vāsana*) is claimed to result from that visit. Thus it was in the processual context of reasoning about the fishes' behaviour (desire to cool) and the sensual properties of the fishes' flesh (smell, taste) that the Lotus Sea was mentioned. Unambiguously, it is part of the *puṟaṃ kaṭal*.

While the *puṟaṃ kaṭal* is visible daily and known to the fishermen from direct experience, knowledge about the Lotus Sea is inferential and reflects assumptions about the behaviour of the fish. (Indeed, fish behaviour is often talked about in terms of human behaviour.) These assumptions are not far-fetched but are derived from the local village context. For example, many small freshwater ponds (*kuḷaṅṅaḷ*, sing. *kuḷaṃ*) in the village are covered with lotus petals. Fishes are seen to be hiding under these petals. Nevertheless, the smell and taste of those fish which are said to have gone to the Lotus Sea are considered to be the strongest argument for its existence. Not only fishermen are of this opinion. I have also heard this from older villagers inland who have no personal relation to fishing. The smell and taste of lotus is thought to have been assimilated by the fishes during their residence in the Lotus Sea. Indeed, in Ayurvedic medicine animals and plants are assumed to assimilate the qualities of their environment (Zimmermann 1987: 7-8; Vasavi 1999: 45). A well-known Malayali saying refers to a similar process of assimilation: 'If a rock lies next to a jasmine bush, it will also smell like jasmine' (*muḷḷappūmboṭijēṭṭu kiṭakkuṃ kallinmuṇṭāmoru saurabhyaṃ*).

What will become clear in the next section is that, in a sense, the Lotus Sea makes the local conceptualisation of fish behaviour autonomous. It locates causes within its borders. By providing a reason for the coldness of the outer sea, it accounts for the seasonal migration of such fish as oil sardines and mackerel.

When first describing the Lotus Sea, Vasu had mentioned that the fish move there in the hot season according to a rule (*niyamaṃ*). In talking about the Lotus Sea, reference is commonly made to a rule or, more often, to the scriptures (*śāstraṅṅaḷ*, sing. *śāstraṃ*). When I tried to pursue this point further and asked which scripture (*śāstraṃ*) was meant, the answers were always evasive. Kumaran Thaykattu (eighty years old), a retired fisherman and middleman, gave a typical reply:

Kumaran Thaykattu: It is said that there are seven seas. This is the Arabian Sea (*aṟabi kaṭal*), the Lotus Sea (*tāmaru kaṭal*) is behind it.

Kumaran was not the only person to mention the Lotus Sea as one of the seven seas. Descriptions of lotuses, including those in lakes or rivers, are an exceedingly popular theme in Indian mythology (Penzer 1968). The idea of seven circular ring seas surrounding the Earth, separated by ring continents, is shared by Puranic and later Buddhist and Jain cosmologies (Kirfel 1968). Only in Jaina cosmology is there something like a lotus sea, the *pushkaroda* ocean, 'named as such since, because of sun and moon, it shines like lotus' (translated from the German after Kirfel 1968: 252–253). But this sea is vastly different from the fishermen's conceptualisation. It is the third of the seven ring seas, filled with a nectar-like fluid and lying far behind two other ring seas and two ring continents (ibid.). It remains doubtful if these traditions are related in any meaningful way to the Lotus Sea as conceptualised by the fishermen in Chamakkala.

To qualify accounts of the Lotus Sea with the reference to a rule or a written source, as Kumaran does, may seem to give them an authority. It might be an attempt to give credibility to a claim, which (as mentioned above) is beyond personal judgement. It may also, on the contrary, express personal scepticism towards an old and possibly outdated concept. Indeed, the reality of the Lotus Sea has been contested, as I shall discuss in Chapter 7.

Heat, Coldness and Currents in the Sea

The notions of 'hot' and 'cold' are central to the local conceptualisation of the sea. 'Heat' – in the following including the pair 'hot' and 'cold' – is not always understood thermometrically (as measurable with a thermometer). Much rather, it may also refer non-thermometrically, but figuratively, to the emotional constitution of a human being or to ideas of purity and pollution.[10] The consideration of whether, and where, the sea is hot or cold is important to the fishermen when they talk about its constitution. Use of it is made not only when the seasonal variation in the sea was described to me, but also in everyday speech when talking about practical aspects of fishing. The direction and strength of winds (sing. *ka<u>r</u><u>r</u>ụ*) and currents (sing. *nīrụ* or *o<u>l</u>ukkụ*) are also important. Both may affect the heat in characteristic ways. I shall first describe the seasonal variation of heat, winds and water currents as understood by Chamakkala fishermen. Thereafter I shall consider shorter-term changes.

When Vasu Kochikkattu was talking about the fishes' behaviour in the hot season, he mentioned the Lotus Sea in the context of experiencing himself the rising pre-monsoonal heat at the beach. He projected his own discomfort with this heat onto the fish in the sea. Due to the heat, Vasu implied, the fish moved to the Lotus Sea in the west to find refreshing shelter from sunshine under the lotus petals. Similarly, most descriptions about seasonal variations in the sea are from the fishes' point of view, as far as can be guessed by the fishermen. I return to this point later in this chapter. For now it is sufficient to note that fish are believed to be sensitive to heat, smell, currents and sound. Fish generally prefer to stay in cool, silent and calm environments where food is available. Most fish are afraid of loneliness. Coolness and availability of food are considered the most important factors for the fishes' well-being.

At the end of the south-west monsoon in early September (mid-*ciṅṅaṃ* in the Kerala calendar), the sea has been thoroughly cooled by the mountain water and the monsoonal rains. Soon the wind, which had come from the south-west during the monsoon, reverses its direction and starts blowing from the east or north-east. In mid-October the north-east monsoon (*tula varśaṃ*) sets in. It is much weaker than the south-west monsoon, brings less rain and lasts only for the *tulaṃ* month (until mid-November). On most days, there are evening thunderstorms accompanied by rainfall. These rains cool the sea further. For the fish it is agreeable anywhere in the sea (but food is not available everywhere). While the rains subside, the wind from the east (*kara ka<u>r</u><u>r</u>ụ*, lit. 'shore wind') continues, but is generally weak. It is thought to contribute to a continual increase in the heat of the sea. There is agreement, however, that the essential cause for heating the sea is the intense daily sunshine (*veyil*) during the hot season (*vēnal*). This rise in (thermometric) heat begins near the surface of the sea. With continuing irradiation it extends to ever deeper layers: the deeper the cooler. Feeling uncomfortable with the rising heat, fish move to colder areas. To the fishermen, a lack of fish in the top layers of the sea is noticeable at the beginning of the hot season. Typically, this can be perceived beginning with the *kuṃbhaṃ* month in the Malayalam calendar (mid-February to mid-March).

When the hot season continues it becomes impossible for the fish to find a cold place in the near-coastal waters. Now the fish 'climb' (lit. *kayaṟuka*) to the western *puṟaṃ kaṭal* where they can find coldness at a greater depth and stay there. At this time, mackerel (*ayila*) and other fish species move to the Lotus Sea for spawning (lit. *perukkuka*). In the course of the hot season, the prevailing wind direction turns from east (*kiḻakkan kaṟṟu*) to north-west (*vadavarata kaṟṟu*). Strong winds from the north-west (*kaccan kaṟṟu*), which typically last for a few days, are common. The big waves coming to the shore along with this wind make it difficult or impossible for the fishermen to go out to the sea.

The increase in heat continues until the beginning of the *iṭavaṃ* month (mid-May), when the pre-monsoon showers begin. A sustained cooling will occur only with the onset of the monsoon, typically around 15 *iṭavaṃ* (1st of June). The sea is understood to be cooled by three causes: a reduction of sunshine due to the cloud cover, rainfall and, most importantly, the mountain water (*mala veḷḷaṃ*) which comes from the mountains through the rivers into the sea. Both the rainwater and the mountain water are thought to be cooling. Rain is believed to cool only the surface layer of the sea (*kaṭalinṟe mukal*). Fish come to the surface to drink this fresh water (*ñān veḷḷaṃ*). A sustained cooling of the sea requires the arrival of the mountain water. As a result, the fish may descend (inf. *iṟaṅṅuka*) to the near-coastal waters:

Kunhiraman Kochikkattu: At this time [interview conducted on 30 June 2000] the heat is decreasing (*cūṭu kuravu*). So when the coldness begins the fish come down here (*tanuppu varumbōl mīn iṟaṅṅi varunaṭṭu ānu*).

It is apparent that the seasonal movement of fish is linked to the seasonal variation of heat and coldness at specific areas in the sea. In an interview conducted in mid-April, in the midst of the hot season, Chandran Kozhissery explains his view of the expected changes:

Chandran Kozhissery: Now the atmosphere (*antarikśaṃ*) is hot. This is known as the summer (*vēnal*). The water in the ponds and the drinking water are all getting hotter at this time. When the upper water (*mēle* *veḷḷaṃ*) is getting hot, the fish move down and they move to the outer sea (*ulla kaṭal*). They like the cold just as we like to sit in an air-conditioned room. They want to live in a cold area. Men also like the cold more than the heat. So the fish not only move to the *puṟaṃ kaṭal,* they also move downwards (*atiyilēkkụ*). When the rainy season (*kālavarśaṃ*) comes then both the bottom and the top of the sea are getting cold, but the bottom will be even colder [than the top]. At the time of rain the bottom will be as cold as ice [sic] so that the fish move up to enjoy the warmth of the sun (*veyil*). Therefore during the rainy season the west (*paṭiññāṟụ*) and the east (*kiḻakkụ*) become the same [of the same heat]. Then the fish are move to the east and move up, since at the bottom it is too cold. Unbearably cold. But today [during the hot season] they have to go down to feel the necessary coldness. Even the fish in the *puṟaṃ kaṭal* will do so.

Like Kunhiraman, Chandran portrays the fish to be free in moving in the sea: they decide according to their comfort where they want to go and are able to get there. It should not be too hot, nor should it be too cold. His description deals only with the variation of heat in different parts of the sea. Several fishermen, middle-aged and older ones, have expressed such a viewpoint.

Unlike Chandran, Vasu Eranpurakkal describes the changes in the sea which are brought along with the monsoon entirely in terms of the change in the currents (sing. *nīrụ*):

Vasu Eranpurakkal: This is the reason why these fish come to the shore: when the mountain water reaches the west (*paṭiññāṟụ*) we may say that the current is pulling (*nīrụ valikkunnụ*). I will say this: *vadarakara nīrụ* means that the water flows from north-east to south-west. *Teṅṅarakada nīrụ* means that the water flows from south-east to north-west. If there is *tekkan nīrụ* it will move from south to north. The *vadakkan nīrụ* moves from north to south. Then a *kara nīrụ* ('shore current') should also happen and pull (inf. *valikkuka*) from east to west. In these currents only the one from the west is lacking. Now for a long time the current from the west should come. This is the time for it [the interview was conducted on 8 July 2000]. But this current cannot reach here, because the mountain water is lacking. But if the mountain water were present in the sea (lit. *paṭiññāṟụ*-side, 'western side') then it moves and stabs (*kuttuka*, lit. 'to stab, sting') and agitates (inf. *iḷakkuka*) and turns so many things around in it. The fish from the western region, of different size, and also the waste things, all of that comes [to the east] along with the water.

Vasu talks about the change in currents but does not mention the seasonal variation of heat in the sea. He asserts that currents come in pairs: a current from north to south is related to a current from south to north. But due to the lack of mountain water the east-west pair of currents is lacking. If present, the mountain water would move from the east to the west and would carry, in its rebound, fish and other things towards the shore. Like Chandran, Vasu explains the changes in the sea in relation to what happens with the fish. Notably, Vasu attributes less agency to the fish: they are carried along with the current and seem not to be able to move where they want to. Both Chandran and Vasu plainly talk about 'the fish', ignoring possible variations in the behaviour of different fish species.

In the many interviews about seasonal change which Suresh and I conducted, the accounts of Chandran and Vasu seem like end points of a continuum. Often, the effects of heat are brought into relation with those of the currents. Vijayan, for example, explains that the seawater is hot and the mountain water is 'cold like *ice*' (op. cit.). By stabbing (inf. *kuttuka*) the lakes (*tāṭakaṅṅaḷ*, sing. *tāṭakaṃ*) in the sea, it carries the fish towards the east. It seems that, while the impact of currents, winds and rain on the 'heat' of the sea is universally acknowledged by the fishermen in Chamakkala, their relative importance is judged individually. The preference for either point of view does not seem to depend on the age of the fisherman.

Besides these seasonal changes, which occur over periods of weeks or even months, variations of shorter duration are of much interest to the fishermen. Seasonally the hotness of the sea is mostly regarded as being influenced by currents. From day to day, the intensity and direction of the wind creates differences in heat. The time of day is also important. During the hot season fish move down during the daytime. Late evenings and early mornings are best suited for fishing, since then the fish are expected to come to the surface. They will do so also at night, but then shoals of fish can be seen only around new moon nights, since only then is the *kamarụ* (bioluminescence) visible. The direction of origin of a wind (*kaṟṟụ*) determines whether it is considered as heating or cooling. East winds (*kara kaṟṟụ*, lit. 'shore wind') are believed to be heating the surface of the sea and are therefore disliked by the fishermen. Western winds (*paṭiññāṟan kaṟṟụ*) are believed to be cooling and are generally welcomed. This is in accordance with the observation that the south-west monsoon (when the winds come from the western sea) cools the sea thermometrically, while it is known that, during the north-east monsoon, the thermometric hotness of the sea begins to rise.

A Physiology of 'Heat'

In the conversation related at the beginning of Chapter 1, Krishnankutti Korupattu described the sea's desire to unite (inf. *yōjikkuka*) with the mountain water, its rage (*kōpaṃ*) when the mountain water arrives, and its subsequent coolness (*taṇuppụ*) and peacefulness (*śāntaṃ*). In conversations concerning the character of the sea these or closely related sexual connotations are commonly made. Their use is not restricted to telling the myth, but pervades much of the fishermen's everyday talk about the sea as well. Velayuthan Kunnattu relates that before the onset of the south-west monsoon and, less so, at full moon, the *prakṛti* is hot (*cūṭụ*). The sea's desire for intercourse (coll. *kaṭalinṟe mati elakkuna*) is only satisfied when it has been pierced or stabbed (inf. *kuṭṭuka*) by the mountain water (*mala veḷḷaṃ*). Kunhiayyappan Kochikkattu explains that the sea and the mountains are having a get-together (*kuṭikālca*), wherein the sea is drinking (inf. *kuṭikkuka*) the mountain water. Kumaran Thaykattu remarks that the sea is drawing or pulling (inf. *valikkuka*) the mountain water in their union.[11]

Krishnankutti opposed two states of the sea: a 'hot' state of movement/shaking (*iḷakkaṃ*), agitation (*kṣōbhaṃ*) or anger (*kōpaṃ*)[12] and a 'cold' state when the sea is cool (*taṇuppụ*), flat (*marggaṃ*) or peaceful (*śāntaṃ*). Oppositions of this kind are common in everyday talk about the state of the sea. They are not restricted to narrations of the myth. Phalgunan Thaykattu opposes the English word *rough* and the Malayalam word for coolness (*taṇuppụ*), while Vijayan Eranpurakkal opposes the heat (*cūṭụ*) of the sea with the coolness (*taṇuppụ*) of the mountain water, which is 'like ice water' (ice-*pōle veḷḷaṃ ānụ*). Generally, before going on a fishing trip it is discussed whether the sea is hot (*cūṭụ*) or cool (*taṇuppụ*).

These utterances suggest an opposition of 'heat, anger, storminess and sexual lust' and 'coldness, peacefulness, flatness and sexual satisfaction'.[13] When the terms 'hot' and 'cold' are utilised in speech the distinction between 'heat' as a thermometric and 'heat' as a figurative, non-thermometric, emotional state is blurred.

If the human body is a 'meeting-place of influences' and has a physiology but no anatomy, as Francis Zimmermann has written about Indian medical texts (1983: 13–15), my material suggests that much the same can be said about the sea. The dynamics of hot and cold, agitation and calmness, can be considered as a physiological variation resembling those associated with the human body. Subjected to certain substantial influences, the sea cannot but heat up or cool down.

This resemblance of the hot-cold polarities in such distinct contexts can be interpreted in two ways: first, as a diffusionist spread of humoralist ideas which are widespread in other settings in Kerala, particularly in medicine and religious symbolism, or, secondly, by taking a cue from the cognitive sciences, as reflecting an innate disposition of the human mind to conceive of hot and cold as a naive theory of humoralism (Zimmermann 2003). I shall consider both cases in turn, beginning with the prevalence of the hot-cold polarity in contexts that are related to the fishermen's life-world while extending beyond it.

The opposition of hot vs. cold is common to Indian and Sri Lankan discourses about the human body. It has been noticed in relation to bodily states, tempers, food, diseases and rituals.[14] In regard to bodily states it may be considered as one of several local discourses, the others being Ayurveda, including the *ṭridōśa* system, the germ theory of disease, racial ideas, heredity/genetics and ideas of differences in the openness and closure of the people of different castes (see Nichter 1992; Osella and Osella 1996: 40f., 1999: 186–187).

The hot vs. cold (or cool) or heating vs. cooling opposition is essential in describing and conceptualising (human) bodily states. A balance is considered desirable, but in general coolness is preferred to hotness. Heat may be positively associated with life and fertility, writes Brenda Beck about her Tamil data, but taken alone it 'can be highly dangerous. It must be focused and controlled in order to become a source of power which humans and superhumans can utilise' (1969: 553). In ritual, heat is associated with blood and is often represented by the colour red. Coolness is usually represented as white, milk typically being regarded as a cooling substance (ibid.: 569).

Women's excess of blood corresponds to an excess of heat, particularly during menstruation (McGilvray 1982: 37). Sexual intercourse is regarded as both heat-caused and heat-generating (Daniel 1984: 189), but results ultimately in cooling (Fuller 1992: 46). Note that, if the arrival of the mountain water from the mountains in the sea is interpreted as sexual intercourse, the heating and subsequent cooling of the sea runs parallel to (and is in accordance with) these ideas about the female human body.

Female menstruation blood has a powerful and dangerous heating influence on the sea. If a menstruating woman enters the beach, the sea's wrath (*kōpaṃ*), noticeable by a thorough roughening and agitation, may only be comparable with the wrath of the goddess (Caldwell 1999: Chapter 3; see also Obeyesekere 1984: 53). Due to the danger associated with this pollution (*aśuddi*) of the sea, menstruating women must not enter the beach area.

Deities are subject to hot and cold states as well, particularly female deities. Fieldwork conducted throughout India has unequivocally shown that the heat of female deities follows a seasonal cycle, reaching its maximum in the pre-monsoonal hot season. Calendrical rituals are performed to cool this heat (Babb 1975, Daniel 1984, Tarabout 1986, Caldwell 1999).[15]

In Araya lineage temples in Chamakkala, such cooling rituals are done at the annual festival in the hot season. These festivals always include offerings of *guruti* (a vegetarian substitute for blood, see Tarabout 1986: 145) for Bhadrakali or Bhagavathi, two forms of the goddess that are central to these temples' pantheons.[16]

Unlike these deities, the sea does not need any human intervention for its seasonal cooling, or at least it did not until human intervention (especially the construction of dams in the mountains) prevented its heat to cool down.

It is tempting to consider the fishermen's concept of the sea's physiology of heat simply as deriving from these pan-regional discourses. However, there is ample evidence for a cross-cultural regularity in the meaning of the hot–cold polarity as well. This is shown by ethnographic material from Latin America to South-East Asia, and it seems rather unlikely that this fact can be explained by diffusionism. Francis Zimmermann (2003) raises the question of whether there is an innate disposition of the mind to conceptualise our environment in these terms. This idea seems plausible in the light of recent findings of the cognitive sciences. Evidence is growing for an innate, inherited human capability to learn language as well as innate conceptual universals whose nature and scope is still debated (see Hirschfeld and Gelman 1994; Astuti et al. 2004, and references therein). Atran (1990, 1998) finds from studies of plant and animal taxonomies that there is an innate recognition of living kinds (a 'folk biology'). In contrast, Carey (1985) finds that there rather is an innate predisposition for psychology, because infants characteristically ascribe intentionality to animated behaviour ('folk psychology'). She claims that 'folk biology' emerges from 'folk psychology' in the course of child development. Zimmermann (2003) draws on Keil (1989), an associate of Carey, in considering concepts such as 'folk biology' and 'folk psychology' as developing from naive theories which are developed progressively during the child's growing up in specific environments. The hot–cold opposition may be one such naive theory of humoralism, surmises Zimmermann (2003: 264).

Let us consider the material on the sea's variable 'heat' from these 'diffusionist' and 'innatist' viewpoints, respectively. The cooling of the sea after its intercourse with the mountain water and the heating influence of menstruating women relate to discourses popular in Kerala and India beyond the fishing community. In contrast, everyday utterances of the sea as hot or cold seem free from other cultural elaborations. As such, these may derive more immediately from an 'innatist' disposition. Even though it is impossible to prove this point with my material, it suggests that the perception of the sea's physiology of 'heat' may derive from two distinct conceptual layers in the fishermen's minds.

Explaining a Mudbank

This discussion of processes involving a physiology of 'heat' is helpful in considering local explanations of the seasonal coming to be and passing away of a mudbank in the sea near the coast. Since this mudbank forms at different places from year to year, it invites discussions about causality in the marine environment.

The seasonal formation of a *cākara* had been observed in the eighteen years prior to my arrival in Chamakkala. Recognisable as a distinct patch of calm, turbid water, it stretches along the coast for about five kilometres and about three kilometres into the sea. It is a calm area in the sea, surrounded by the rough monsoonal swells. Known from several places along the coast of Kerala, this phenomenon is called a *cākara* (from *cākụ*, death, and *kara*, shore). Though usually translated into English as 'mudbank' (e.g. Kurup 1972: 1; Silas 1984), a Malayalam–English dictionary translates *cākara* as 'occurrence of mud-banks and appearance of plenty of fish for a short period at some places along the Kerala coast during the monsoon', and adds the idiomatic meaning of *cākara* as a 'season of plenty, plentifulness' (Warrier et al. 1999: 369). This is no surprise, since locally the *cākara* is of great importance as the only place where the fishermen can enter the sea during the rough swells of the south-west monsoon (which is the season in which the greatest catches are usually made). Occurrences of *cākara*-s are often linked (wrongly) to large catches of (costly) prawns. *Cākara*-s typically form at the beginning of the south-west monsoon (June) and dissolve at its end (August/September).

The formation of *cākara*-s during the south-west monsoon has been reported from several places along the coast of Kerala in between Kozhikodu and Kollam. In this region the sandy sea bottom gradually falls off towards the west. Most *cākara*-s are semicircular, extending about five to seven kilometres along the shore and three to four kilometres into the sea (MacPherson and Kurup 1981: 154). Kurup (1972: 3–4) lists nineteen locations where *cākara*-s formed between 1835 and 1971. While some of Kerala's *cākara*-s are seen at the same place each year, others are known to shift to different locations irregularly (Gopinathan and Qasim 1974: 105). The *cākara* near Chamakkala is of the shifting type: it is said to have come here from a village about five kilometres in the north, but is supposed to have originated about fifty years ago near Chavakkadu, a place about thirty kilometres to the north. It is an open resource to all fishermen. Fishermen from Chamakkala and the adjacent villages agree that anybody could come from anywhere to fish here, provided they behave honestly (see Chapter 5). Two small surveys I did of the boats fishing in the *cākara*-s of 1999 and 2000 yielded that these came from up to fifty kilometres away.

In 1999, more than 200 boats used the *cākara* near Chamakkala to enter the sea or as a fishing ground. In addition, there was a large number of men fishing with cast nets (*vīśu vala*) and stake nets (*vaṭi vala*). The visiting men belonged not only to the Hindu fishing castes (Araya, Mukkuvar and Valan), but also to the Izhava and Vettuva castes, as well as to the Muslim and Christian communities. Such uncontrolled open access to fishing is not practised at all of Kerala's *cākara*-s. Some of the adjacent *cākara*-s in the north and south, e.g. Ponnani (Malappuram District) and Purakkad (Allapuzha District), are controlled by local committees. Both of these *cākara*-s are fixed to one place. In this way, they can be subjected to local control more easily.

Vishwambhara Panikker, an astrologer and *mantravādi* (lit. 'expert in mantras, magic spells') who regularly works for the fishermen, knows of a myth about the formation of the *cākara*. It deals with Lord Shiva and his sister *Kaṭalamma*, here considered to be a form of Devi, the goddess:

Vishwambhara Panikker: There is a plant [*koduvēli*] which changes things into gold once they are touched by it. It is said that it comes from the mountains and is used for meditation (*yogaṃ*). Devi uses it to scrub (inf. *tēcukaḻukkuka*) and to bathe (inf. *kuḷikkuka*). They [the fishermen] call Devi by the name *Kaṭalamma*. Devi is the sister of Lord Shiva. He is sending these things (*sādanaṅṅaḷ*, sing. *sādanaṃ*) [the *koduvēli* plant] to his sister. That is the concept. It comes here along with the mountain water (*mala veḷḷaṃ*). Just like we are sending presents on Vishu and Onam to our sisters. In this way this thing is sent.

[Once it is in the sea] in these things the *cākara* is there. It is said so according to one story. In that area the water is not moving at all. It [the *cākara*] stays in one place and then it comes closer and closer [to the shore]. Fishes and other things are present in it. It is considered as the *caiṭanyaṃ* (life force, consciousness) of Devi. It is said that it is the Devi. It is related to Devi.

Here the *cākara* is seen as the manifestation of divine consciousness or presence (*caiṭanyaṃ*; the same word is used for the deities' presence in a temple), substantially transmitted by the mythical *koduvēli* herb. The *cākara* here appears to be the by-product of a gift made by Lord Shiva to his sister Devi/*Kaṭalamma*. This is not an unlikely constellation in Indian myth: a strong bond between siblings is a classical theme in Hindu mythology and life. The giving of gifts by brothers to their sisters at the Onam festival has been documented, for example by Caroline and Filippo Osella (1996: 52). A similar custom prevails locally on Vishu, the beginning of the astrological and agricultural year, celebrated on the first day of the *mēṭaṃ* month (mid-April). Lord Shiva's residence in the Western Ghat mountain range may be understood in terms of the attribution of male gender to these mountains. The magical strength of the *koduvēli* plant can be traced to a homology between the Himalayas (wherein Mount Kailash is known as as his favourite place for meditation) and the Malayas (the Western Ghats), which is occasionally emphasised, especially with regard to the efficacy of herbal medicine originating there (Zimmermann 1987: 70).

In this myth the *cākara* is made by the intentional action of a divine agent (Lord Shiva) extraneous to the sea. It is not made by the sea itself. Locally, another account relating how the *cākara* is made intentionally by extraneous agents is known. It is the story of two Thangals, high-status Muslims often associated with knowledge of *mantravādaṃ* (sorcery), who were talking to each other while sitting on the beach near Allapuzha (in Southern Kerala, the place of the most famous *cākara*). They felt disturbed in their conversation by the noise of the incoming waves and told the sea to 'shut up'. Allegedly, the sea understood this, behaved and calmed down. The first *cākara* was formed.

Chandran considers such accounts with scepticism or disbelief. Now in his late sixties, he started as a fisherman on his father's *vañci* and eventually became a *tarakan* (middleman). He explains:

Chandran Kozhissery: It is said that the blue (*nēla*) *koduvēli* and about 465 other plants move from the mountains with the mountain water and when they reach the sea then the *cākara* is made by these plants. A plant named *pala* and another one named *veḷḷikoduvēli*. In the olden days such a story was heard, but I don't believe it. In my opinion if it is happening like that then it [the *cākara*] should not move away from that place. In my experience (*anubhavaṃ*) the *cākara* is formed when the monsoon is happening and thus the sea gets rough and because of the rough sea, the strong wind and big waves and big tides and so many other things are happening and when these things are biggest then the *cākara* forms. That is my experience (*anubhavaṃ*). If the *nēla koduvēli* has an important role in it or if it happens by the wisdom of God (*iśvaradīnaṃ*) then the *cākara* should not move away from there. The *prakṛti śakti* of this *cākara* lasts about ten to twelve years. Everybody says so and I also feel that this is right. Thus for twelve years it exists in one area and after that it moves further south.

Unlike Vishwambhara, Chandran does not refer at all to Lord Shiva or *Kaṭalamma* and their *caiṭanyaṃ*. Instead, he ponders over the efficacy of the herbs to make the *cākara*, an idea of which he strongly disproves. He implicitly considers a cleansed version of the myth, stripped to an empirically verifiable core, and judges this in the light of his own experience (*anubhavaṃ*). In doing so he realises that the myth is inconsistent with the observed movement of the *cākara*. According to the myth it should always form at the same place near a river mouth. This contradicts the *cākara*'s observed southward motion. Similarly, he asserts that it would not move if it were due to the wisdom of God (*iśvaradīnaṃ*). Instead he feels that a combination of several factors may lead to the formation of the *cākara*.

Other men are not so sure whether some special grasses (*pratyēkata pullụ*) coming into the sea with the mountain water may not at least contribute to the flattening of the sea. Like Chandran they understand the origin of the cåkara in the agitation of the sea in combination with a local accumulation of mud. Velayuthan Nediripil, in his mid-forties and a shareholder on a *vaḷḷaṃ*, gives a vivid account:

Götz Hoeppe: Why is the *cākara* so calm (*śāntamāyi*)?

Velayuthan Nediripil: That? That is coming ... mud (*cērụ*) is the thing. Mud (*ceḷi*) is most important (*pradānaṃ*) for it. If we think in which area the sea is not breaking (*oṭiyuka*), then we notice, first, that the waves are not breaking where it is deeper. Secondly, the waves are not breaking where there is a lot of mud (*ceḷi adhikamulla stalaṭṭaṃ*). The waves are not breaking at the time when the mud is standing (*nilkkunaṭṭụ*). Even if the sea is rough (lit. 'becoming more and more', *kuṭutal kuṭutal*) the waves will not break there [in the *cākara*]. Similarly, butter is formed when milk is agitated. In the same manner, at the time of the agitation [of the sea] (*iḷaki varuna samayaṭṭụ*), it [the calming of the sea] has to happen.

While Chandran had been content to relate his experience, Velayuthan tries to make the *cākara* be understood by comparing it with a more familiar process. The analogy of the churning of milk to make butter and the formation of the *cākara* is capable of relating the latter's calmness to the agitation of the sea. A comparison with the Rig Veda's milk ocean, churned by the Devas and Asuras to yield *amrita*, the nectar of immortality, may spring to mind. However, Velayuthan does not seem to refer to mythology, but rather suggests an analogy from his rural surroundings. Like most other men, Chandran and Velayuthan agree that mud is of constituting significance in the formation of the *cākara*. They add that the sea must be rough for this mud to become stirred up and note that the *cākara*'s seasonal formation in June or July is likely to be related to the monsoonal rain (*maḻa*), the roughening of the sea (*kaṭal iḷakkunnụ*) and the arrival of the mountain water (*mala veḷḷaṃ*). Its disappearance with the onset of the eastern wind (*kara kaṟṟụ*) in August/September is also noted as a regular event, as well as its being present at a place for only twelve years. As such it is often termed a *prakṛtiyuṭe śakti*, a 'power of the *prakṛti*' or a *prakṛtiyuṭe svattụ*, a 'property of the *prakṛti*'.[17] Most of these explanations involve mud (*cella or cēṟu*), which is either thought to have arrived in the sea with the mountain water or may have been previously accumulated on the sea floor near the beach. With the roughness of the sea or the arrival of the mountain water this mud will be stirred up (*cēṟu iḷakkunnu*). When it has joined or united (inf. *khaṭṭikuka*) it will fall (inf. *kaḻiññuka*) at a certain place and lie there (inf. *kiṭakkuka*). This is the *cākara*. It will remain at this place until it is pushed to the west by the shore wind, which commences at the end of the monsoon. Many people insist that despite this regularity some special power (*pratyēkata śakti*) or divine power (*daivaṭinṟe śakti*) must necessarily be involved in the formation of this phenomenon, particularly when its formation at a particular place is debated.

Let us now turn from explanatory accounts of *cākara* formation to everyday utterances made on the beach near Chamakkala in 2000, when the formation of a *cākara* was hoped for. In late June, a few weeks after the south-west monsoon had commenced, the *cākara* was expected to form. Since March, boats and nets had been overhauled, new nets made, engines bought, storage huts repaired. The air was full of hopeful expectation for the coming season. In May discussions began revolving around reports that the signs (*lakśyaṅṅaḷ*, sing. *lakśyaṃ*) of a *cākara* were seen in the sea. This talk was often like talk about an expected but little-known visitor. Whenever its position or movement was debated, whenever the 'behaviour' of the *cākara* was considered, its agency and intentionality were emphasised. People talked about its birth (*utbhavaṃ*) and, once born, at which place in the shore it would reside (inf. *tamasikkuka*). Later it was considered to be on the 'look-out for a comfortable place to lie down (*kiṭakkānulla saukaryamānu pulli nōkkunatu*)'. The Malayalam pronoun *pulli*, often used when talking about the *cākara*, addresses a familiar male (third person singular) person. Some places were regarded as more convenient (lit. *saukaryamulla*) for the *cākara*'s residence than others. But this year it seemed rather undecided throughout the season. A local Muslim *mantravādi*, whom the *cākara* had always obeyed (inf. *kēlkkuka*), was called to the beach. But the outcome of his rituals was judged with ambiguity. It seemed that other *mantravādi*-s were pulling (inf. *valikkuka*) on it from different sides. While one *cākara* had initially formed it seemed to have split into two parts by mid-July. Some claimed that it had been cut (inf. *kaśnikkuka* or inf. *marikkuka*) by the use of *mantraṃ*-s.

It is generally agreed that the *cākara* has become increasingly unsteady in recent years: 'Today the *cākara* moves with the *mantravādi*-s ['sorcerers']' or 'Nowadays the *cākara* is [behaves] like a politician (*cākara ippōḻatte rāśṭriyakāreppōle*)'. Both the works of sorcerers and the doings of politicians are unpredictable, unreliable and can be influenced by money (see Chapter 7).

Non-mixing and Mixing Waters

Explanatory accounts of *cākara* formation have added the influence of substances on the sea's physiology of 'heat', a theme that is further elaborated in local ideas about the formation of fish.

Towards the end of the south-west monsoon, when rain is followed by days of intense sunshine, *pola* is commonly observed in the sea. These are extended patches of pink-coloured water of a slimy and sticky consistency. These patches are noticeable by their intense putrefactive smell and by causing, once approached, a burning sensation in the eyes. For the 'fish' (*mīn*), *pola* is considered harmful. Says Phalgunan Thaykattu: 'Fish entangled in the *pola* have got the behaviour of an intoxicated person (*viśaṃ kaḻippa ālkaruṭe mātiri*)'. They try to escape, but, being unable to control themselves, often get stuck in it and die. Even though the occurrence of *pola* is restricted to the surface of the sea, it is most dangerous for bottom-dwelling and slow species. Fish like *cāla* (oil sardine) and *ayila* (mackerel) are believed to be smarter and are rarely found dying in the *pola*. There are no benefits from *pola*: 'It is only a *dōśaṃ* (fault)', says one man. But what is the *pola*? 'It is like something dead (*oru catta-pole ānu*)', says another. The common explanation of the *pola* considers fresh water (*ñān veḷḷaṃ*) reaching the sea as rain or via the rivers without mixing with the sea's salt water (*uppu veḷḷaṃ*). It remains near the surface. Due to the warmth of the sunshine (*veyil*) the water begins to rot (inf. *ciyuka*) and turns into *pola*.

Vasu Kochikkattu compares the formation of *pola* with a situation of indigestion (from the sea's point of view):

Vasu Kochikkattu: That [fresh] water cannot be joined (*yōjya alla*) with the seawater, just as if we are taking food which is not suitable for our stomach then what happens? Somehow the body should discard it and only after that do we feel better. If we take unsuitable food (*paṭātta bhakśanaṃ*) then it does not join to our body. It may decay and somehow get out from the body. The same happens to the rainwater when it falls into the sea. At the time of decay it should have the red colour ... If any fish comes into the *pola* then it will perish. In the *pola* they cannot open their eyes, so they are becoming blind. If they drink the *pola* water they die ... If we are inhaling toxic oxygen what will happen to us? Or if we are eating poisonous food what happens? Then we are in danger! Just like that the seawater is the [natural] medium of the fish and if they get near the *pola* they will die, because *pola* is not good for them. If they eat *pola* they are intoxicated (*tala karaṅṅiyiṭṭu*, lit. 'head turning') and come up unconscious.

Thus, fresh water does not mix with the salt water, and what is initially attractive to the fish may turn into a toxic and deadly substance.[18] These explanations about the formation of *pola* are remarkably similar to the ones recorded by James Hornell, then Government Marine Biologist in Madras, when he visited northern Malabar in November 1908 (Hornell 1931: 54; *pola* seems to be what he called *karanir*, shore water or shore current).

In other instances, rainwater is assumed to mix with the salt water. It is believed that this mixing results in the formation of *tari* (small fragment, seed). This *tari* is understood to be the food of prawns (*cemmīn*), which, while usually staying near the sea bottom, come to the surface to eat it. But not all men agree with this claim. Others say that fish feed on some unspecified 'garbage' (*tāra*) in the water or on mud or some substance dissolved in the mud.[19] Many older fishermen accept *tari* as the food of prawns and other marine species. Younger fishermen are often reluctant to give their own answer. They are sceptical of old beliefs but often unable to state a consistent alternative. My questions about the food of fish were often found surprising. Apparently, not much thought is invested in this question.

Jellyfish (coll. *coḻi*) are believed to form in a similar way to *pola*. Krishnankutti Korupattu explains:

Krishnankutti Korupattu: This pure water moves to the sea and it stays in the outer sea (*puṟaṃ kaṭal*) without mingling with the salt water. Then it becomes rigid like an ice block, then it should change into small pieces like this [he shows his fist]. Then in its face, hands and legs form and it gets life (*jīvaṃ*) also. The things like ice pieces which have hands, legs and life are known as the jellyfish (*coḻi*).

While *pola* is believed to be a dead substance, other men agree with Krishnankutti that, somehow, (living) jellyfish form from almost the same process. That jellyfish expand and contract, that they move in the sea and that they sting is proof of their being alive. But even though jellyfish are considered *jīva jālaṅṅaḷ* ('living water-like things') they are not considered *mīn* ('fish').

'Fish'

A pertinent question thus arises: What is *mīn* ('fish')? 'What lives in the water and can be eaten', says Subramoniam Kovil Tekku Valappil with a smile (Figure 4.1). This is not generally true, since *kākka* (molluscs) and *ñaṇḍu* (crab) also live in the water and can be eaten, but are not called *mīn* ('fish', see Table 4.1). There is a general agreement that *cāla*, *ayila*, *cūṭa* are

Figure 4.1: Vijayan Kovil Tekku Valappil (left) and Subramoniam Kovil Tekku Valappil (right) pondering over the anatomy of fish

mīn, and so are *sravụ* (shark) and *tiraṇṭi* (stingray). *Cemmīn* (prawns) are always considered *mīn*, not the least because of its name (lit. 'red fish'). It was also agreed, that *cāla* (oil sardine) and *ayila* (mackerel), which together with *cemmīn* make up most of the catch of the *vaḷḷaṃ*-s, are more typical as *mīn* than *cemmīn*. By lacking fins, scales etc. the latter lack some of the features which most *mīn* ('fish') share. As such, *cemmīn* is noted to be anomalous. More than any other fish, *cāla* (oil sardine) may be considered as the prototype of fish, as far as its anatomy and behaviour are concerned. This is not by chance: the largest fraction of the catch is *cāla*. On the other hand, crab (*ñaṇḍụ*), jellyfish (*coḷi*) and sea snake (*kaṭal pāmbụ*) are never called 'fish' (*mīn*). Of these, only *ñaṇḍụ* is eaten. What is clear from these findings is that the semantic fields of Malayalam '*mīn*' and English 'fish' are not identical. In fact, the semantic field of *mīn* is more inclusive than that of 'fish'. While acknowledging this distinction, in the following I shall drop the quotation marks of 'fish', unless I want to draw attention to the distinct semantic fields of these terms.

Chamakkala fishermen agree that fish are living things or beings (*jīvajālaṅṅaḷ*). Moreover, and unlike the sea, fish are perceived to share anatomical traits with humans. In describing the anatomy of fish (head, mouth, eyes, spine, body) fishermen use the respective Malayalam terms describing humans. All fish are considered to have perceptual skills and emotional preferences similar to those of humans. They are capable of

Table 4.1 Cross-identification of the names of important marine species.

'Fish' ***(mīn)***

avoli	Stomateidae spp.	pomfrets
ayila	Rastrelliger kanagurta	Indian mackerel
cāla	Sardinella longiceps (Val.)	Indian oil sardine
cūṭa	Kowala coval	white sardine
ēṭi	(unidentified)	porpoise or dolphin
eṭṭa	Tachysurus spp.	cat fish
kanambụ/tiruta	Mugil cephalus	grey mullet
kora	Sciaenidae spp.	jew fishes
malan	(unidentified)	(unidentified)
manaṅṅụ	Engraulidae spp.	anchovies
mantal	Cynoglossus semifasciatus	Malabar sole
mullan	Leiognathidae spp.	silver bellies
parava	Lactarius lactarius	big-jawed jumper
sravụ	Carcharhinidinae spp.	sharks
tiraṇṭi	Mobulidae spp. Mulibatidae spp. Tryngomadae spp. Rhinopteridae spp.	rays
valamuri	(unidentified)	a sole-like, bottom-dwelling species
velōri	(unidentified)	(unidentified)

kaḻuṇtan cemmīn	Metapenaeus affinis
karikkadi cemmīn	Parapenaeopsis stylifera
ñāran cemmīn	Penaeus indicus
pōvalen cemmīn	Penaeus dobsoni

'Living things [in the water]' (jīva sadānaṅṅaḷ) that are not 'fish' (mīn)

coḻi	jellyfish
kakka	eatable mussels found near the beach
kaṭal āna	whale, lit. 'sea elephant'
kaṭal pāmbụ	sea snake
ñaṇḍụ	crabs

Sources: Ramachandran Nair (1986: 237–245), Kumaran et al. (1988) and Bal and Rao (1990).

Note: The Malayalam names of several fish species used in south Kerala and central Kerala differ. For example, the Indian oil sardine (*Sardinella longiceps*) is called *mati* or *nalla mati* in Thiruvananthapuram District, but *cāla* in Trichur District. The table lists the names commonly used in Thrissur District.

hearing, sight, taste and sensing hotness and coldness. These capacities are unequally distributed among fish species. For instance, *valamuri* (ribbonfish) cannot bear the heat and therefore prefers to stay in the mud at the sea bottom, *ayila* has better sight than *cūṭa* and *cāla* is considered smart while *cūṭa* is regarded as stupid. Fish may be happy or afraid and most prefer other fishes' company instead of remaining alone. They also prefer calm surroundings rather than a turbulent sea. As such, their preferences are similar to those of the fishermen. Oftentimes, however, 'fish' are plainly referred to as 'substance' (*sādanaṃ*), 'goods' / 'commodity' (*caṟakkụ*) or 'matter' / 'stuff' (*vastụ*), especially (but not only) in practical contexts of fishing. This stands in marked contrast to the elaboration of the sea and the *cākara* as social others.

Needless to say, knowledge about fish and their behaviour is inferential, being mostly derived from the practice of fishing and inspecting the catch. Many situations in the process of fishing invite speculations about fish behaviour: the composition of a catch may indicate the co-sociality of different species, the place and time of the catch of a certain species is a hint at its migration patterns and the visible behaviour of near-surface schools can be directly watched. The above examples suggest that in Chamakkala fish behaviour is modelled on ideas of human capacities and preferences. Some fish are born from eggs (*muṭṭụ*), but others are born as 'children' (pl. *kuṭṭikaḷ*). That they gather and travel in schools (*pelappụ* or *kūṭṭaṃ*) confirms their nature as social beings. The analogies between humans and fish extend much further than in the case of the sea: so far that, when Chandran Kozhissery explains that in fishing 'the living ones are moving to kill the other living ones' he puts fish and humans into one category (albeit a fairly general one: that of 'living ones', *jīvanullavan*). Fishing is commonly compared with hunting (*nāyāṭṭụ*), suggesting that an awareness of the fishes' habits and preferences is important for success in fishing. A fisherman has to know that fish are sensitive to sound. This can be used to chase (inf. *celluka*) the fish by beating on the boats with wooden sticks, as was formerly done. In other instances it needs to be remembered that noise (e.g. of engines) may chase fish away.

The uncertainties in trying to predict fish behaviour are known all too well. Says Kunhiraman Kochikkattu: 'These are the fish. It is not their habit to build a house and live in it. So they may travel from one side to another'. Indeed, if any place or space can be called the home (*parkkuna stalaṃ*) of the fish, then it is the outer sea (*puṟaṃ kaṭal*). The fishermen are familiar with fluctuations in the catch of fish, particularly of *cāla*. These may disappear from the nearby sea for weeks, months or even years and are then said to have migrated to other parts of the Arabian Sea (*arabi kaṭal*) or the Bay of Bengal (*bengal ulkaṭal*). The fishermen's uncertainty about the sardine's migratory habits pertains to other fish species as well:

Kunhiayyappan Kochikkattu: The *prakṛti* (nature, character) of the sea's fish cannot be explained by us. Sometimes the fish steps down [comes to the shore] and then the heat is not at all a problem for them. The *prakṛti* of the sea, or the *prakṛti* of the fish in the sea; it means how the *prakṛti* of the sky is changing in the same manner the sea and the fish are changing. Our ancestors mentioned three conditions for it: the rain from the sky, the fish in the sea and the changes of the *prakṛti* of the sea, these three cannot be predicted by anyone and nobody can do anything about it.

Thus, while general ideas about the behaviour of fish can be formulated, the fishermen are aware of the fact that some matters are unknowable by humans.

Much of what Chamakkala fishermen know of different species (sing. *vamśaṃ* or *ellataraṃ*) can be characterised by a set of oppositions, most of which relate to fish behaviour. There are (a) fish which spawn vs. those which give birth to their children, (b) those which form schools vs. those which live alone, (c) those which can move faster than the currents vs. those which cannot and (d) those which live in the upper water layers vs. those which live near the sea bottom. In order to illustrate the fishermen's conceptualisation of fish, oppositions (a), (b) and (d) can be depicted as the three axes of a cube (Figure 4.2).

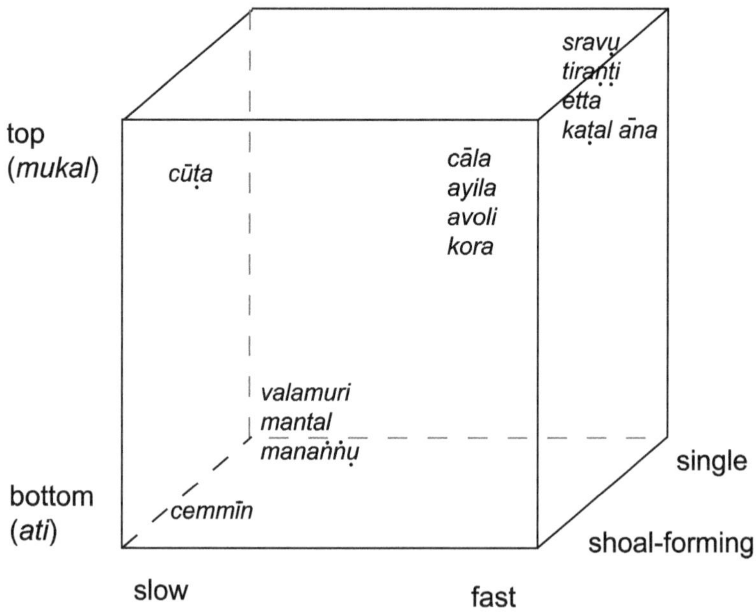

Figure 4.2: Fish species are distinguished by their behaviour

(a) Most fish reproduce (inf. *perukkuka*, 'to multiply') by spawning (*paṭṭal*, lit. 'sticking, adhesing'; inf. *paṭṭuka*, 'to let fly', or inf. *muṭṭayiṭuka*, 'to lay eggs'). Exceptions are shark (*sravụ*) and rays (*tiraṇdi*), who give birth to their 'children' (pl. *kuṭṭikal*) by delivery (*prasavaṃ*). Among spawning fish, oil sardine (*cāla*) and white sardine (*cūṭa*) are believed to lay many eggs, while anchovies (*manaññụ*) lay few eggs and are said to lack interest in intercourse. This is one of the reasons why species like anchovies (*manaññụ*) are believed to be particulary affected by trawling. Moreover, the eggs of *cāla* and *ayila* are said to float in the water while those of the fish of the *varśapeytụ* ('monsoon catch') are laid at the bottom and likely to be damaged by bottom trawling.

(b) The majority of fish are known to form shoals (*pelappụ* or *kūṭṭaṃ*), notably oil sardines (*cāla*), mackerel (*ayila*), prawns (*cemmīn*), white sardines (*cūṭa*), silver bellies (*mullan*) and pomfrets (*avoli*). A shoal may include fish of different species, e.g. oil sardines (*cāla*) and mackerel (*ayila*). Rare species like *pūlan* are known to join the schools of other fish. The supposed reason for forming shoals is fear of loneliness and for safety from predators. Species like shark (*sravụ*) and stingrays (*tiraṇdi*), believed to reside near the bottom, occasionally form small groups, particularly when near the surface.

(c) Whether or not fish can move faster than the currents determines their agency (i.e. their capability to act intentionally). Those who can are able to choose quiet and cool places according to their convenience and move around in search of food (*bhakśanaṃ*). Examples are *cāla, ayila, avoli, kōra, eṭṭa, sravụ* and *tiraṇdi*. It is assumed that *cāla* is the fastest fish. The enormous fluctuations in the catch seem to prove this assumption. Fast fish are able to escape from the engine noise that is caused, for example, by trawling ships.

Bottom-dwelling fish are generally believed to be less fast than the currents. Examples are *valamuri, manaññụ, mantal, velōri* and *cemmīn*. Among the fish that are found near the surface, *cūṭa* and *parava* are believed to be slow. The monsoon catch (*varśapeytụ*, lit. 'rain catch') consists of these slow fish. They are believed to be carried to the shore with the current from the west. It is this current that is occasionally surmised to be the counter-current of mountain water. In this sense the lack of the mountain water is interpreted to be the reason for a lack of 'slow' fish near the beach during the south-west monsoon.

(d) The distinction between those fish which live near the surface of the sea and those living near the bottom is largely reducible to the distinction between school-forming fish and loners (b). Most school-forming fish live near the surface, while single fish usually stay down at the bottom.[20] A counter-example is *cemmīn* which forms shoals whether it stays at the bottom or moves to the surface.

These oppositions distinguish fish in a simple, but revealing way: *cāla, ayila* and *avoli* are characterised as spawning, school-forming, fast fish which usually stay near the surface of the sea. Prawns (*cemmīn*) spawn, form shoals, are slow and stay, usually, at the bottom. White sardines (*cūṭa*) spawn, form shoals, are slow and usually stay near the surface of the sea. *Valamuri* spawns, is mostly single, is slow and stays at the bottom of the sea. Shark (*sravụ*) and stingray (*tiraṇṭi*) are born by delivery, are sometimes single, sometimes in groups, fast and usually at the bottom (but occasionally near the surface). In Figure 4.2, I have attempted to represent graphically the order that is implied by these oppositions.

To classify fish along these lines provides the fishermen with an understanding of the regularities of their behaviour and differences among species of fish. It is also useful when considering the effectiveness of the fishing techniques that are used to catch different species: shoal-forming fish seen near the surface are the target of ring-seine nets: *cāla, ayila, avoli* and *cūṭa*. Bottom-dwelling fish, which cannot be seen from the surface, would be caught near the coast from *vañci*-s with *nītụ vala* (gill nets), or, recently, with trawling and double nets (e.g. *valamuri, mantal, cemmīn*). Single fish in the invisible part of the sea, including the bottom, may be caught with hook and line (*cūṇda*).[21] The scheme is also the basis for judging the effect of recently introduced gear on the fish stocks. Trawling, for instance, is generally believed to affect only bottom-dwelling species like *cemmīn* and the fish of the *varśapeytụ* (monsoon catch; see section (a) in the classification scheme mentioned above). It is not believed to affect the catch of *ayila* and *cāla*.

Forming shoals is considered to be a measure against loneliness, but also a precaution against predators: some fish are known to feed on other fish. Among the species known as predators are: *arkia, ēṭi, sravụ* and *kaṭalāna*.[22] These feed mostly on *cāla*. Other commonly known examples of predation are *kōra* feeding on *mantal* and *kuduta* feeding on *cūṭa*. Vasu Eranpurakkal answered my question about the food of fish like this: 'The big fish eat the small fish and the small fish also have some food'.

Cemmīn

Of the species that constitute the catch in Chamakkala, prawns (*cemmīn*) are anomalous in several respects. First, they lack anatomical features (e.g. fins and scales) that are regarded as defining most fish morphologically. Secondly, their seasonal migration between the sea and brackish water regions near the shore (e.g. the backwaters) is regular in comparison with the erratic migrations of other species. In addition, the catch of prawns is much more rewarding economically than that of any other species. Prawns are India's most important seafood export item, sold mainly to Japan, the European Union and the United States, and a large fraction of this catch is secured by Kerala's artisanal fishermen. In Chamakkala, past catches of prawns worth six, seven or eight lakhs (600,000 to 800,000 Rupees) by local *vaḷḷaṃ* units are remembered and occasionally talked about.[23] These are large amounts of money suited to pay off debts, but being exceedingly rare they also allude to a wealth obtainable from the sea that generally remains imaginary.

Locally, the most important prawn species is *ñāran cemmīn* (Penaeus indicus), followed by *pōvalen cemmīn* (Penaeus dobsoni), *kaḻuṇtan cemmīn* (Metapenaeus affinis) and *karikkadi cemmīn* (Parapenaeopsis stylifera). It is generally known that all but one of these prawn species migrate seasonally between the sea in the west and the backwaters in the east.[24] The only species that remains in the sea during its entire lifespan is *karikkadi cemmīn*. The migratory species are known to lay their eggs in the western sea and remain there until the beginning of the south-west monsoon.[25] At the beginning of this monsoon, the young prawns move to the backwaters and return to the western sea at the end of the monsoon.[26] Prawns are known to form shoals (*pelappu*) when migrating. The eggs (sing. *muṭṭu*), laid in the hot season (October until March / April), are variously described as garland-like or as resembling the shape of light bulbs, but in both cases they are said to be in or near the mud (*cēṟu*) at the sea bottom. In describing further stages in the development, the appearance of the prawns is usually not talked about much. Ratherm, it is the size development (relevant for the economic benefit of a catch; larger exemplars sell for a better price) that is known. Regarding *ñāran cemmīn*, small exemplars may be caught in the sea from the Malayalam *mēṭaṃ* month (mid-April to mid-May) until the beginning of the south-west monsoon (15 *iṭavaṃ*, about 1 June), while larger exemplars are seen in the Malayalam *mitunaṃ* and *karkkaṭakaṃ* months (mid-June to mid-August). It is commonly thought that the lifespan of *ñāran cemmīn* is about thirteen months.[27]

While statements about the size development of prawns are fairly similar, opinions about the reason for the prawns' seasonal migration differ considerably. According to Kunhiraman Kochikkattu, prawns migrate to the east at the beginning of the south-west monsoon because of being attracted by the coldness which they may experience in the near-shore waters (the *arayi*). This resembles the general assumptions about the seasonal preferences of fish. Vasu Kochikkattu argues that the prawns come to the near-shore waters during the south-west monsoon because this is where they will find food:

Vasu Kochikkattu: For the fish to come down from the *paṭiññāṟu* (west) the heat is a must. This fish comes near the shore and stays under the

> water. The monsoon rain forms a *tari* in the water and the *cemmīn* comes up to eat it. When they come up we run to the boats. Thus the *cemmīn* first comes up to the water [surface].

Thus, according to Vasu prawns may only be caught when feeding on *tari*, since otherwise they stay at the sea bottom. I discussed *tari*, an alleged product of mixing rain and seawater, as the food of fish at the beginning of this chapter. Some fishermen doubt that prawns feed on *tari* and rather contend that they feed on the mud at the sea bottom.[28] Asked about the food of prawns, Kunhiraman Kochikkattu responded:

Kunhiraman Kochikkattu: They [the prawns] are different from us. It may be the mud (*cēru̥*); except for the mud I don't see any food for the *cemmīn*. In some places the mud is special (*pratyēkana*). I think it is the mud and the water itself, except for that nothing is there.

While fish like *cāla*, *ayila* or *sravu̥* are not known to suffer from any disease, *cemmīn* and crabs (sing. *ñaṇḍu̥*) may suffer from fever (*pani*). Prawns of all the species mentioned above are believed to suffer from fever at full moon (*velutta vāvu̥*) and at new moon (*karutta vāvu̥*). At these times some, but not all, prawns lose their shells (sing. *tōṭṭu̥*). Kunhiraman Kochikkattu explains that, when suffering from fever, 'the prawns get soft and hang to one side or another, like a tuberculosis patient'. Yet this fever is not fatal. It is recognised as a disease because fish merchants do not buy prawns that have lost their shells, explains Kunhiraman.

Many fishermen in Chamakkala are aware of the fact that, because of their laying eggs at the sea bottom and their migration to the backwaters, prawns are endangered from two types of fishing that are not practised in the village itself. First, it is known that trawling means to drag a net over the sea bottom and thus damage the eggs of prawns. Secondly, backwater fishermen are known to catch prawns upon their return to the sea with stake nets and so-called Chinese fishing nets (*cīna vala*; operating with fish-attracting lights at night). The local fishery of prawns is encroached from the east as well as the west.

Interpreting Figurative Language

The fishermen's speech seems to abound with instances of figuration, whether the men talk about the sea as mother and as female or refer to its constitution with terms similarly used in speaking about the human body or its physiological/humoral processes. All of these expressions can be termed anthropomorphisms, but doing so must not distract us from their variations in meaning and the manifold possibilities of their interpretation.

Interpreting such utterances one has to choose either an insider's viewpoint or the viewpoint of an external observer. Choosing the former, one may claim that these examples do not belong to figurative language (and metaphor as its most prominent form) at all. While metaphor may ordinarily be defined as 'understanding and experiencing one kind of thing in terms of another' (Lakoff and Johnson 1980: 5), a phenomenological viewpoint would draw on personal experience, in which the two realms basic to a projection are not clearly distinguishable from each other. Michael Jackson (1989: 135) notices the 'thisness of that' in the interpretation of metaphor. He speaks of a 'mode of thinking, speaking and acting in which personal, social and natural aspects of Being are made to correspond and coalesce' (ibid.: 137). Choosing the viewpoint of an external observer, I distinguish the fishermen's figurative language from what seems to me as their understanding of 'the literal'. Philosophically and linguistically, this is an awkward position (Davidson 1979, Rorty 1987, Hesse 1993), and I am led to wonder whose metaphors I am studying, the fishermen's or my own? Both viewpoints are laden with difficulty of interpretation, and any statement about the distinction of literality and figurativeness must by necessity be a relative statement. That Chamakkala fishermen themselves do have a concept of the metaphorical is most evident when one considers instances of shifting meaning and the men's growing critique of notions such as the rise of the sea towards the west or the Lotus Sea (see Chapter 7).

Max Black suggests that 'every metaphor is the tip of a submerged model' (1979: 31). This idea is useful independently of the awkwardness involved in drawing a line between the literal and the figurative. Are 'the sea is mother', the physiology of 'heat' and the myth of the intercourse of the mountain water retold in Chapter 1 different tips of one and the same underlying anthropomorphic model? It is impossible to explore this possibility with my material, because I am unable to determine the extent to which 'hot' and 'cold' mirror a naive theory of humoralism.

Chamakkala fishermen do perceive talking of the sea as mother or referring to the physiology of 'heat' as strange (implying an awareness of the figurative), yet they also know that these terms capture both the regularities and the irregularities that are experienced in the marine environment. They score well in terms of accuracy, consistency, scope, simplicity and fruitfulness.[29] Talking about the sea in terms of its 'standing' or 'lying down', its 'anger' or 'calmness' makes successful communication possible, and, even if the underlying anthropomorphic concepts are not shared by all men (as I show in Chapter 7), such expressions can be thought of as an 'observation language' that is 'theory laden' (to borrow two terms from the philosophy of science) by its reference to an underlying anthropomorphic model. The same applies to talk about the *cākara*, in which anthropomorphic expressions that shorten longer descriptions of observable facts abound – and are agreed upon far more than the myths in which this anthropomorphism is transformed into literal fact. Conceiving of the figurative in such terms means to shift the reference of metaphor from individual words (as in classical rhetorics) and parts of sentences (as in Black's interaction view) to the context of communication (see Ricoeur 1985, Sperber and Wilson 1995).

This said, instances of figurative language only reveal one side of the local conceptualisation of the sea. It must be supplemented by a consideration of the fishermen's social relatedness to the sea and its ritual enactment, and I shall turn to this in the following chapter.

Notes

1. Besides the popular version of the above-mentioned myth in which the sea is female and the mountains (or the mountain water) are male, there is an alternative version in which the sea is female (daughter) and the rivers are female as well (mother). Vasu Eranpurakkal explains it like this:

   Vasu Eranpurakkal: Whether the sea is a lady and the mountains are a lady or a man: that is not the matter here. You know: the mountain water comes from the east. The two are mother and daughter, it is said like that. This [the sea] is the daughter and that is the mother. The daughter wants to see the mother and the mother wants to see the daughter. They stay separated (*nilkkunna oru̥ avastajillu̥*). Then the daughter said: 'I come to meet you, mother!' But the mother said: 'You cannot visit me. If you come the whole world will be lost. I will come to see you!' That is what we call the mountain water (*mala veḷḷam*).

   As in the myth presented at the beginning of Chapter 1, to protect humanity living in the coastal stretch is a conscious decision, in this case made by the (female) mountain water. In yet another version, the sea was identified as the mother and the rivers as children: when the children come to the mother they would embrace (inf. *koṭṭippiṭikkuka*) and roll (inf. *mariyuka*).
2. These two pilgrimages draw hundreds of thousands of pilgrims each year from all over South India (and beyond). The pilgrimage to the shrine of Lord Ayyappan at Sabarimala has been described, for example, by Daniel (1984) and Fuller (1992). The shrine of Lord Murukan at Palani has been studied by Moreno and Marriott (1990). A legend says that the mountain on which Palani is situated, was brought from the Himalayas to Tamil Nadu by the demon Idumban (Eck 1998: 172).
3. Examples from the English-language press are 'Thakazhi, the man and littérateur' (*The Hindu*, Kerala edition, 11 April 1999, p. 4) or 'A man who wrote language of life' (*The New Indian Express*, Thiruvananthapuram edition, 11 April 1999, p. 5).
4. Note that Pillai does not identify *kaṭalamma* with the sea itself, but describes the sea as the abode of *kaṭalamma*.
5. Many fishermen (especially young or middle-aged ones) consider fishing as a (low-prestige) physical labour (*pani*), much less valued than permanent work in a shop, in an

office or as a teacher (*jōli*). On the distinction between *pani* and *jōli*, see Osella and Osella (2000: 71).

6. English words are increasingly entering the vocabulary of Malayalis, whether they speak English or not.
7. While *bhagaṃ* denotes depth, horizontal distance is measured in *maṛu*. Like one *bhagaṃ*, one *maṛu* is 1.7 metres long.
8. Houses in Kerala are usually built with an elevated floor to prevent rainwater from entering.
9. When Hindu fishermen in Chamakkala refer to God they allude to the concept of *iśṭadevata* or *sēvāmūrtti*, the personal god chosen by a worshipper (Fuller 1992: 268). For fishermen, this is often a form of Devi, usually Bhagavathi or Bhadrakali, but may sometimes be Ayyappan or Subramoniam.
10. In other Indian discourses it has been noted similarly that 'hot' and 'cold' do not always refer to a scale of thermometric temperature. Ice cream, for instance, is regarded as very hot (Daniel 1984: 185). One must therefore not assume that 'hot' and 'cold' are related to thermometric temperature in a definite way. On heat as a metaphor of nervousness, anger, wrath, etc., see Lakoff (1987: 380–415).
11. According to Dossi (1998: 20–25), in Vedic literature rain was considered as semen. Given that mountain water is known to be accumulated rainwater, it is reasonable to assume that the mountain water could have been understood as an equivalent of semen.
12. For Suresh and me, in the speech of some men, especially older ones, the distinction between *kṣōbhaṃ* and *kōpaṃ* was often difficult to make, even on tape recordings. With regard to the sea, the former denotes (literally) 'agitation', while the latter denotes (figuratively) 'anger'. The difference between these words is slight, but potentially meaningful. It may signify minor 'plays of tropes' (Fernandez 1988) or cases of 'slippage' (Uchiyamada 1995: 124–127).
13. Similar connotations have been identified in a very different cultural context (late twentieth century North America) by Lakoff (1987: 380–415).
14. Examples from the literature are Beck (1969), Babb (1975), McGilvray (1982), Daniel (1984), Obeyesekere (1984), Nichter (1992), Osella (1993) as well as Osella and Osella (1996, 2000).
15. An example for a cooling ritual from Central Kerala is the subject of Caldwell's (1999) study: *muṭiyeṭṭu* (lit. 'hair dance'). Traditionally performed on a dry, harvested rice field (ibid.: 107) in the dry season (between mid-December and mid-May; ibid.: 114), *muṭiyeṭṭu* is a ritual enactment of Kali's slaughtering of the demon king Darika. It is meant to cool the wrath of Bhadrakali, a fierce form of the goddess. A multicoloured image of Bhadrakali (*bhadrakali kalaṃ*) is drawn on the ground in colourful organic powders and then destroyed by a person possessed by the deity. When correctly done, it results in cooling the goddess and fertilising the earth with blood for the upcoming cool monsoon season. During the time of my fieldwork in Chamakkala, several *muṭiyeṭṭu* were performed in inland villages nearby, but none was done in Chamakkala itself.
16. During my fieldwork Araya lineage temple festivals in Chamakkala and its neighbouring villages were held between January and March, i.e. before the peak hot season. Though differing in elaboration, these festivals were similar in conduct. The *guruti* offering was always done during the last night of the festival, after a day of (more or less) elaborate processions, in a ritual enactment of the battle of Kurukshetra. Thereafter the temple was closed for a period of eleven days.
17. The Sanskrit term *prakṛti*, which Chamakkala fishermen sometimes use, has often been translated as 'nature' (e.g. by Vasavi 1999: 29). As Francis Zimmermann emphasises, *prakṛti* does also refer to 'constitution' in a Hippocratic medical sense (1983: 13). It is a manifestation of divine female power (Gold and Gujar 2002: 8). Gold and Gujar note the difference in the semantic fields of 'nature' and *prakṛti* (ibid.). They retain the term

'nature' in their book, yet not for 'accuracy but for ambiguity, complexity, and uncertainty' (ibid.: 11).

18. Fish are said to come to the surface after a rain shower to drink the fresh water because of its good taste. Bhimachar and George (1950) have studied the 'red water phenomenon' along the Malabar coast. It seems to be identical with *pola* and is ascribed to a bloom of *Noctiluca* phytoplankton organisms.
19. That fish feed on mud was related to Hornell (1910: 97) as early as 1908 by fishermen from northern Malabar.
20. Some fishermen say that the fishes' place of residence can be inferred from their anatomy: *cāla* and *ayila* have their mouths at the centre and are believed to live in the middle to upper parts of the sea. In contrast, supposedly bottom-dwelling species like mantal, *sravụ* and *tiraṇṭi* have their mouths at the bottom.
21. Nobody in Chamakkala is still doing hook-and-line fishing. The only man who did hook-and-line fishing until recently was Hamsa from Kaipamangalam, who is locally known as the *cūṇdakkāran* (the 'hook-and-line man'). Hamsa told me that the decline in the catch of sharks has made this type of fishing redundant.
22. The whale (*kaṭalāna*, lit. 'sea elephant') is not always considered a fish, it is also called *timiṅṅalaṃ* (*ti* being a Malayalam prefix for sacred beings or things). There are other land-based animals that seem to have equivalents in the sea: *kaṭal panni* ('sea pig', dolphin), *kaṭalāna* ('sea turtle') and *kaṭal pāmbụ* ('sea snake'). Kumaran Eranpurakkal once told me that all that is present on the land can also be found in the sea. This may appear as a 'parallel world' to the one on the land. The parallel, however, does not extend much beyond these animals and the surmised presence of a few topographical features (hills and lakes) in the sea bottom.
23. These numbers may be compared with the Ponnonam unit's annual catch worth about 40 lakh (4,000,000) rupees in 1999. One lakh rupees were about 2,500 euro in the year 2000.
24. The prawn species migrating between the sea and the (brackish water) backwaters are: *ñāran cemmīn*, *pōvalen cemmīn* and *kaḻuṇtan cemmīn*.
25. In scientific parlance, eggs are laid in the sea by adult prawns and pass the nauplius, protozoa and mysis stages of (larval) development. Then they enter the backwaters in their post-larval stage and develop into juveniles. As such they return to the sea (George 1988).
26. Consequently, much of the fishing activity is concentrated near the waterways connecting the backwaters with the sea. Unlike the other prawn species, *karikkadi cemmīn* lays its eggs in the near-shore area and migrates towards the western sea during the south-west monsoon (see Suseelan et al. 1989).
27. This number, which is approximately correct (Bal and Rao 1990: 330), probably stems from an extraneous source of information.
28. Bal and Rao (1990: 330) note that prawns feed on nutrients in the mud.
29. These are the five 'values' (in common usage: standards) which the philosopher of science Thomas Kuhn (1977: 321–322) proposes to compare different scientific theories.

5 Socialising the Sea

Chamakkala fishermen occasionally compare fishing with hunting (*nāyāṭṭu*). Abounding in dangers and potential failure, both aim at the killing (*samhāraṃ*) of life. From the viewpoint of a Brahmanical ideology, this poses a moral dilemma. Hunting and fishing are depreciated activities that stand in opposition to the ideal of non-violence towards living beings (*ahimsa*). But what about the fishermen's ideology? The material presented in the last chapter suggests that the answer may be found in the fishermen's perception of relating to the marine environment socially. But how is this sociality constituted? In this chapter I explore two modes. One is the direct interaction between the fishermen and the sea, the other operates through the mediation of ritual experts and deities. In this context the enquiry must then turn towards what it means to kill fish.

Truthfulness

Kunhiraman Kochikkattu: That [the sea] is something special. It is truthful and righteous. It is a valuable and good thing, isn't it? That western side, oh! It is what protects us!
(*atu oru pratyēka sādanam ānu itu satyanītiyulla mutalānu nalla vastu, allē? paṭiññāṟuppurattatu, ēi, nammale kakkunagu gunallē!*).

Every fisherman in Chamakkala is familiar with the concept of the 'from the beginning truthful sea' (*satyamulla mutal kaṭal*).[1] A Malayalam-English dictionary translates *satyaṃ*, a word of Sanskrit origin, as 'true, real; truth, especially of a promise; an oath, an ordeal, truly indeed (adv.)' (Gundert 1995: 1031, see also Söhnen-Thieme 1995). For the fishermen, the sea's truthfulness (*satyaṃ*) is a normative concept that subsumes the notions of purity, protection and provision and emphasises reciprocity. Altogether, it defines a 'code of conduct' which, though in principle it can be practised by everyone, is held to be characteristic of the people residing in the

Figure 5.1: Chamakkala fishermen deliver a golden ornament (tali) to Palappetti Bhagavathi.

beach area and distinguishes them from the people of the outside world, particularly those living in the 'outside in the east', the *kiḻakkuppuṟaṃ* (see Chapter 2). The notion of truthfulness is shared by Hindus and Muslims who are involved in fishing and fish vending (and who typically reside in the beach area). While the term *satyaṃ* is unknown to Brahmins and Nayars of the inland in the contexts described below, it subsumes practices and understandings known to Kerala Hindus more generally. Truthfulness can be considered as a local religiosity that is precipitated by the sea as the coastal dwellers' shared social other. It helps to differentiate between the 'westerners' and the 'easterners' and is part of a discourse that counters the hegemony of the east.

Purity

The sea is considered as innately truthful, but truthfulness is not exclusive to it: 'Everything in the universe (*lōkaṃ*) has got the *satyaṃ*', says Kumaran Eranpurakkal, 'everything that has been created by God (*daivaṃ*)'. In this generality, *satyaṃ* may seem rather unspecific. However, it turns out that only the sea is properly truthful:

Kunhiraman Kochikkattu: Mountains, forests and hills may be truthful (*satyamulla*), but with these things of the east (*kiḻakkattụ karyaṅṅaḷ*) one can never be sure if they really are. There may be pure ones (*vrittiyullavan*) and impure ones (*vrittiyiḻḻattavan*). There is certainly nothing in the east whose truthfulness (*satyaṃ*) and punctuality (*kṛtyaṃ*) can be compared to that of the sea.

In general, places inhabited by human beings are not considered to be truthful, but temples, sacred groves (sing. *kāvụ*) and shrines of Brahmin spirits (sing. *brahmarakṣassụ*) are (the latter being much respected by local Muslims as well). In this regard, *satyaṃ* is synonymous with ritual purity. Evidence for the truthfulness of *kāvụ* and *brahmarakṣassụ*, both usually situated on privately owned land, is provided by their contribution to the prosperity (*aiśvariaṃ*) of a household and its members. Some plants are considered to be truthful (*satyamulla*): *aruta*,[2] *plāvụ*[3] and *cetti koduvēli*.[4] Similarly, the *cākara* is considered to be innately truthful.

In these cases, truthfulness is synonymous with ritual purity and the sacredness of certain plants or places. The association of truthfulness with purity (*śuddhi*) extends beyond the domain of ritual. Polluting the sea must be avoided. Theft, murder, omitting offerings and the use of (black) sorcery are regarded as moral impurities. Impurities may be physical, or rather substantial, as well. In particular, the female body is considered a serious source of pollution (*pula*,[5] *aśuddhi*), especially during menstruation (*hṛtụ*). Hindu women may visit the beach (*kaṭappuraṃ*), but they must not do so during menstruation. If that happened, some older men warn, the seashore would perish (*kaṭappuṟaṃ naśikkụ pōkuṃ*). The sea would be provoked to become rough (*kaṭal iḷakkuṃ*) and be full of anger and wrath (*kaṭalinṟe kōpaṃ*).

During my fieldwork I have seen Araya women at the beach only on very rare occasions, usually drying fish or, once, with the newly wed husband for an evening stroll. When once an accident happened near the beach, Araya women came only to the nearest coconut palms near the beach to watch; they did not enter the beach proper. Older women agree that it is improper for women to visit the beach:

Radha Kovil Tekku Valappil: The sea does not like it (*kaṭal iśṭamilla*). It is said that the sea is becoming angry (*kaṭalinṟe kṣōbhaṃ varuṃ*). I feel that it is true. What our ancestors say is true.

Likewise, local Muslim women are rarely seen on the beach. Muslim or Christian women from the east (*kiḻakkuppuṟaṃ*) are seen more often, always accompanied by their husband and family. To them the sea and the *cākara* are destinations for sightseeing.[6] In recent years, several Tamil women have been working in unloading fish from the *vaḷḷaṃ*-s during the *cākara* season. In doing so they step into the water. Older fishermen comment on this with displeasure. In contrast to the older ones, younger fishermen are often sceptical of whether female pollution can affect the sea.

The *cākara* is said to be particularly prone to pollution. Women must not come anywhere near the beach in the proximity of the *cākara*. If a menstruating woman does so, the *cākara* might move away. Some men claim that a *cākara* which had settled near Chamakkala a century ago disappeared for this reason.[7]

A woman is considered to be polluting for a period of seven days. During this time she should seclude herself from the family (sleeping on the veranda outside the house) and must not participate in religious activities, including visits to temples, sacred groves or *brahmarakṣassụ*. She should not prepare food during this time, but this rule is less strictly followed. Nowadays, the full period of seven days is rarely observed. In comparison with their women, the men's capacity to pollute the sea is believed to be limited. Before going fishing they must wash their body if they had intercourse with their wives, otherwise it is said that an accident might happen. It is not considered good to spit, urinate or defecate into the sea, but due to the length of the fishing trips this could not always be avoided. If necessary, this was to be done on the side of the boat opposite to the net (*puṟaṭṭụ pakkụ*, lit. 'outer side'). It is not said to cause serious pollution of the sea. Other sources of pollution are birth and death. Birth pollution lasts for sixteen days if the newborn is a son and for twelve days if the newborn is a girl. At the death of a family member, the pollution lasts for eleven days.[8] During this time all close relatives must abstain from sexual intercourse, eating meat and fish and religious activities. The lineage temples are closed during these periods. However, there is no prohibition on going fishing or selling the catch.

Protection and Provision

When I asked fishermen in Chamakkala what the concept of the sea's truthfulness implied, most of them referred to its character of providing food and of protecting the fishermen:

Balan Irinjathuruththy: *Satyaṃ* means that we move to a bigger thing. When we move, we can see the big waves. But these waves do not hit us.

The protective character of the sea is also made explicit by the myth of the intimate meeting between the sea and the mountains (see Chapter 1) and by the rising of the sea towards the west which is explained as being due to the sea's *satyaṃ* or the protective might (*kaḻivụ*) of God. Lastly, providing and protecting attributes may be subsumed as nurturing and 'motherly', and, indeed, the concept of *kaṭalamma* is often mentioned in relation to the sea as being truthful.

Reciprocity

The sea is believed to respond to impurities and moral misconduct. These are considered as truthless (*asatyamulla*) and are punished by the sea in a punitive but rational way. As such, truthfulness can be understood as a reciprocal relationship between the sea and the fishermen. In a conversation with Gangadharan, Vijayan Eranpurakkal eloquently elaborates on this point and reflects about the changes that have occurred in recent years:

Vijayan Eranpurakkal: Today people work for their personal benefit only. Some people are still truthful (*asatyamulla*) and righteous (*nētiyulla*), but only those who work in the coastal area (*tēradēśaṃ*). Here you can keep lakhs of rupees[9] while you are sleeping. Will anybody kill you by beating [in order to steal the money]? You cannot tell me a single story that this has happened! If a man sleeps on the beach we may think: let the poor man sleep after his hard work.

Gangadharan Kochikkattu: Your bag of money will be kept where it is. Nobody will open it.

Vijayan: And what is the reason for that? Because they are the people who move to the sea for a living! *Kaṭalamma* has *satyaṃ*. And in the name of *satyaṃ* these things [i.e. stealing, beating, killing] must not be done. The sea is endless (*aṭṭamiḻḻatu*). If we look towards the west (*paṭiññāṟu*) we cannot see the end. We move to it for a living, and because of that we also have the *satyaṃ* and are righteous (*nētiyulla*). In the thought of *kaṭalamma* we must not disturb anybody. We move to *kaṭalamma*, so we need to show our *satyaṃ* here. If we do not do so then *kaṭalamma* may enquire about it. We are afraid of that.

Gangadharan: If we do, then the same thing should happen to us.

Vijayan: There is another thing about the *satyaṃ*. We return with fish [from the sea] to the *cākara* area. [On arriving] here we offer so many fishes to the beggars and other ones. Assume that we caught fish for 1,000 rupees. From that we certainly give 200 rupees' worth of fish away [as charity]. You see this Ashokan? He is a [disabled] man who cannot work. He can also take the fish. Do the people of the east (lit. *kiḻakkuppuṟaṃ*) give even one rupee to others [as charity]? If a man tells them that he is hungry they will not give him money. But the fishermen will give. For all these things the fear (*bhayapāṭu*) towards the mother is the important matter [reason]...

There is an old proverb (*colḷu*) like this: *Kaṭalamma* lies there spread out (*kaṭalamma iṅṅanu vistarāmayi kiṭakkunnu*). We only take 50 or 100 *maṟu* of net. There are so many other ways through which the fish can escape. So we get it because of the *satyaṃ*. And we have to pay some of it as charity. That is in our mind.

Gangadharan: It is an old proverb (*colḷu*).

Needless to say, theft and murder happen on the seashore as anywhere else. However, Vijayan and Gangadharan emphasise the normative character of truthful behaviour, which, ideally, would precipitate a moral order in the coastal area and thus distinguish it from the east.

Truthfulness must be practised not only individually, but also collectively, for example by a *vaḷḷaṃ* unit. Once a year, on the *vaḷḷaṃ*'s birthday, the Ponnonam team distributes ten sacks of rice to the poor of the village. Offerings are given to up to fourteen temples, churches and mosques. The lineage temple run by the team's astrologer and *mantravādi* receives offerings almost daily. These are given in cash money and partly constitute the astrologer's salary. Monthly offerings are brought to the temple of Palappetti Bhagavathi, their *para devata*. Irregularly, but several times a year, offerings are brought to the Guruvayur Krishna and Triprayar Rama temples. Annually, a *tali* (golden ornament) is offered to Palappetti Bhagavathi (see below) and small *tali*-s are brought to the Aaranmula Bhagavan temple, a temple of Parthasarathi, the 'deity of vehicle conductors'.[10] Once a year, several members from the team participate in the pilgrimage to Sabarimala in the mountains of southern Kerala. Before they leave for the pilgrimage a *pūja* is done in the name of the team. Several team members participate in the pilgrimage to Palani. Depending on astrological consultations (*prasnaṃ*-s) offerings to other temples may be appropriate, e.g. Chottanikara Devi. Muslim and Christian members of the team bring offerings to mosques and churches. The expenses are paid by the team.

In mid-June 2000, two weeks after the onset of the south-west monsoon, an accident happened to the team's *vala vaḷḷaṃ* while it was being navigated to the beach by several young men. The damage was estimated at 350,000 rupees (about 7,500 euro). After the accident, older fishermen blamed the young men for causing the accident, accusing them of being out for fun and lacking respect towards the sea by being truthless (*asatyamulla*). In yet other discussions, the mechanical cause of the accident was debated. The next day, these interpretations were supplemented by a consultation with the deity Vishnumaya Chattan,[11] who was invoked in a possession ritual in an Araya lineage temple in Chamakkala. The consultation was done in public as part of a regular monthly festival at the temple whose priest is the astrologer and *mantravādi* for the Ponnonam team. This led to the conclusion that one of the young men had omitted to bring an offering (*vaḻipaṭu*) to a temple near Ernakulam (sixty kilometres south of Chamakkala) in the name of the team, an incident of truthless (*asatyaṃ*) behaviour. To make up for their wrongdoing, the team offered a *tali* to the Chottanikkara temple. It was delivered by three men from the team, including the culprit, who did penitence at the temple.

Offering at Palappetti

In the notion of truthfulness the fishermen interact with the sea directly. In contrast, offering at a temple involves the priest's mediation through his ritual conduct. According to a popular saying, only half of the success in fishing is due to one's own action. The other half is attributed to God's doing (*bhagati nammal bhagati daivaṃ*, lit. 'half we half god'). Bringing offerings (*vaḻipaṭu*) to temples means maintaining the relationships to specific deities. It is part of the fishermen's 'technology of production' (Gell 1988).

The idea of the sea as a 'mother', who provides food but is punitive in a rational way, elaborated above, needs to be supplemented by a closer look at the fisherfolk's relation to Palappetti Bhagavathi, the 'local deity' (*stala devata*) of the beach and their 'chosen deity' (*iśta devata*).

The Palappetti Bhagavathi temple is run by a committee of (high-status caste) Nayars, its priest (*pūjari*) is a Nambudiri Brahmin. The relationship between the fishermen and the goddess of Palappetti is made visible by the *bhanḍāraṃ* box installed at the beach in Chamakkala, the monthly *vilakkṳ*, when the temple's sanctum is illuminated by oil lamps (paid for by the Ponnonam team), and the annual procession of delivering a golden ornament (*tali*). Flowers blessed in the temple and delivered by the priest as part of the *prasādaṃ* (sanctified food or other substances distributed to worshippers at the end of *pūja*) are sometimes taken to sea on a fishing trip and thrown into the water. These flowers are said to contain the deity's *caiṭanyaṃ* (life force, consciousness).

There are other large Hindu temples nearby, but Palappetti is the most important temple of the goddess in the vicinity. In the temple, her image (*vigrahaṃ*) is installed facing west. This is rather unusual for a temple of the goddess, whose images are usually installed facing east or north (Induchoodan 1969: 249–250; Beck 1976). People from all communities know that local Hindu fishermen claim to have had close ties with the Bhagavathi of Palappetti for a long time. A booklet published by the Palappetti temple committee explains: 'The deity faces west. This is different from other Devi temples. It is said that she faces the sea in the west in order to witness 1001 *samhāraṃ*-s (killings). Once this act is executed the deity is appeased by the "children of the sea" (*kaṭalinṟe makkalākayāḷ*). The compassionate deity blesses them always. This is their belief' (Purushooththaman 1995: no pagination, translated from Malayalam). In this understanding, killing fish is made an act of sacrifice to the goddess, who is not cooled through a priest's ritual (as is the case of most female deities, who are cooled in seasonal rituals; see Chapter 4) but through the fishermen's killing of life. While Bhagavathi is commonly perceived as the benevolent form of the deity in Kerala, its transformatory character, comprising its ferocious form more commonly associated with Bhadrakali, is not uncommon (Tarabout 1986). Nevertheless, this is only one aspect of Palappetti Bhagavathi, since in many instances its benevolent role of making the sea productive is emphasised.

By offering at Palappetti the fishermen maintain their relationship with the deity through which they are protected and provided with their catch. The annual festival is of central importance. Traditionally, fishermen from Chamakkala and its adjacent *kaṭavu* held a joint procession on the Revati day of the *kumbhaṃ* month of the Kollam calendar (in 2000 in early March), with the fisherfolk (men and women) starting from each *kaṭavu* 's *bhanḍāraṃ* box to the temple.[12] Meeting there they entered the temple together, circumambulated the inner temple compound three times before entering the sanctum (Figure 5.1) and delivered the *tali* to the *pūjari*, who conducted a *pūja*, in which he decorated the image of the deity with it, upon which the fisherfolk were blessed by her *darśanaṃ*, the deity's divine sight.

The annual festival at Palappetti is held for ten days, Revati being the second-last day. The fisherfolk's involvement late in the festival is not by chance, for their entrance (and that of other *avarna* castes) into the sanctum marks the temple's pollution, a necessary part of the festival. Only after being closed for eleven days and ritually cleansed with *pūja* is it re-opened. The involvement of fishermen in polluting a temple of the goddess during its annual festival is also known from the Bhadrakali temple at Kodungallur, fifteen kilometres to the south (see Tarabout 1986: 375). This festival is celebrated in the *kumbhaṃ* month, and thus in the hot season as well. Both festivals are meant to cool the deity (see Chapter 4).

Until the Temple Entry Act of 1936, the procession on the Revati day of the *kumbhaṃ* month was the only time in the year when the fisherfolk could receive their deity's *darśanaṃ* and could be the direct recipient of her bringing fortune, well-being and spiritual merit. This was the only day in the year when the fishermen could bypass Brahmanical superiority in worship. On all other days they had to rely on the mediating role of the priest in their offering to the deity and in receiving her blessing.

Minor Rituals

While offering at Palappetti is done for the collective benefit and protection of the entire beach (as made visible by the *bhanḍāraṃ* box), as well as for the individual benefit of individual teams, further attempts to influence human–environment relations symbolically with ritual are made, often aiming at individual success. These always involve the mediation of a ritual expert and imply a distinction between the sea and the deities who command over it. When exploring the notion of *kaṭalamma* (see Chapter 4), I noticed an ambiguity of the sea being identified with the goddess of the Hindu pantheon. Whenever a priest's action is described in relation to fishing, this ambiguity is resolved towards the sea being a distinct productive resource that is controlled by divine power. Fishermen usually know the purpose rather than the details of rituals, whose structure is, in general, not exclusive to fishing. Many of the ritualists belong to other castes and work for clients from non-fishing communities as well.

During my fieldwork, resort to ritual techniques in the fisheries was an everyday phenomenon. It was most visible by the presence of *yantraṃ*-s in the boats (copper plates with engraved magical diagrams; see Namputhirippattu 1994). These were periodically recharged in a temple by ritual specialists, and then said to contain a deity's *śakti* (power) or *caiṭanyaṃ* (life force, consciousness). These may be Brahmanical deities, like Shiva, Vishnu, forms of Devi and Ayyappan, or little deities, such as forms of Chattan. Such *yantraṃ*-s are meant to exert an influence on success in fishing and to assure the protection of the fishermen. As mentioned above, flowers blessed in the temple are taken along as well, and are thrown into the water for success in fishing.

In general, the purpose of resorting to ritual techniques is to get rid of *dōśaṃ*-s, the latter being defined as a 'fault, defect' (Gundert 1995: 514) or as a 'flaw, harm, fault, sin, bad manners, illness' (Warrier et al. 1999: 538). I shall define all ritual acts that are meant to achieve difficult-to-obtain goals or to remove *dōśaṃ*-s as *mantravādaṃ* ('magic'). Besides offering at Palappetti, this is another part of the fishermen's 'technology of production', as made clear by the participation of a *vaḷḷaṃ*-'s ritual expert in the division of the auctioned money after a fishing trip (see Chapter 3).

There are two subcategories of *mantravādaṃ*: (benevolent) *satmantravādaṃ* and (malevolent) *durmantravādaṃ*. The former involves the worship of Brahmanical deities who are vegetarian and non-violent. In Chamakkala, such *mantravādaṃ* is regarded as truthful (*satyamulla*). In general, it is not kept secret but is conducted in public for an entire beach (*kaṭavu*) and is of collective benefit, like the offering brought to the Palappetti Bhagavati temple. The latter involves the propitiation of non-vegetarian, violent, little deities, including Chattan, to whom offerings of blood and alcohol are given. In Chamakkala, this is generally regarded as truthless (*asatyamulla*) and is done secretly, either by the ritual expert himself or with the attendance of a small group of clients. If mutual accusations of resort to ritual techniques are made (as occasionally happens), one usually has acts of *durmantravādaṃ* in mind. It is used primarily to benefit individual fishermen or teams, and is the subject of rumour. Besides the propitiation of little deities, *durmantravādaṃ* may also include the symbolic manipulation of objects. Offerings into a ritual fire (*homaṃ*) may be done in either kind of *mantravādaṃ*. Both of these involve the recitation of magical spells (*mantraṃ*-s), usually in Sanskrit. Ritual acts that include the recitation of *mantraṃ*-s and the performance of a *pūja* are called *karmaṃ*-s.

A ritual expert is a *karmi* or *mantravādi* (lit. 'expert in *mantraṃ*-s'), there being a considerable overlap in his activities with those of an astrologer (*jyoltsyan*) and a priest (*pūjari*). Often these are one and the same person. In reasoning on the efficiency of their work, ritual experts often refer to their personal relationship with a little deity, which they regularly propitiate by offerings (*vaḻipaṭu-s*) and can ask for particular favours (see Tarabout 1993).[13]

Influencing the Cākara

Talk about the formation of the *cākara* often contains references to ritual techniques being resorted to with the intention of attracting the *cākara* to a specific beach and ensure its 'residence' there (see Chapter 4). Manoharan (pseudonym) is one of the experts involved. He gives a typical account of how the experts' action is understood locally:

Manoharan: People from all beaches (*kaṭavu-s*) work hard to get it [the *cākara*]. With *mantraṃ*-s and *karmaṃ*-s everybody pulls the *cākara* to their shore. If it comes to a *kaṭavu* then everybody believes that God has blessed these people. And it is true that the *cākara* is a blessing of God. I can explain to you what the *cākara* is. When the sea in the west (*paṭiññāṟu*) is strongly agitated then the mud from below the water [i.e. the sea bottom] comes up. The waves always move towards the east and thus along with the waves the mud reaches the shore. That mud settles at the seashore. And because of this mud the water cannot break at this area. So it becomes calm there. At this time, whatever dirt or garbage or whatever is present in the water comes with the mud and settles at the shore. In my mind, that is the *cākara*. People do so many intense *karmaṃ*-s in order to get the *cākara* to their own shore. Humans are not sacrificed, but all other sacrifices (sing. *bali*) are made.

Suresh Muttichoor: What is the connection between sacrifice and the *cākara*?

Manoharan: By doing the sacrifices they are pleasing the Gods. By doing the *karmaṃ*-s I and you and everybody can make the Gods happy. Brahmins are not needed for that. The *duśkarmaṃ*-s are mostly done by Parayans and Pulayars [low-status castes of agricultural labourers]. The beach people bring these *karmi*-s to the seashore and the *karmaṃ*-s are held there. They pull it [the *cākara*] to the shore by means of the *karmaṃ*-s. That is the attraction, just like a magnet attracts iron. There is a particular *mantraṃ* to do this attraction, listen carefully!

(He recites a *mantraṃ* in Sanskrit, it remains incomprehensible to Suresh and me.)

In the same manner I can attract you as well. In this manner we attract the *cākara* by using these kinds of powerful *mantraṃ*-s. If I make the god so happy then the whole mercy of the god should be with me to fulfil the *karmaṃ*. What kind of bali has to be done with these *karmaṃ*-s is well described in the texts. We do the *homaṃ*-s [offerings into a fire] and bali-s first to please the God. If it was an effective *karmaṃ* then we will see good results.

In order to prevent our *karmaṃ*-s from being effective, people from other *kaṭavu*-s do anti-*karmaṃ*-s. Those *karmaṃ*-s are known to bind [inf. *baṇdhikkuka*, to tie]. Almost all *bhaṇdhānaṃ*-s are done by Hanuman. If it is done by Hanuman then nobody [else] should get it. In order to release the *bhaṇdhāraṃ* we have to approach Sri Rama.[14] If one is using a *mantraṃ* then the other resorts to a more powerful one. Thus it is going on.

(He shows the text of the Hanuman *bhaṇdhānaṃ mantraṃ*.)

With this *mantraṃ* we pray to Hanuman in order to do the binding (*bhaṇdhānaṃ*). Then he does it. Then the offerings are done for Hanuman and one asks Hanuman to encircle the things with his tail. If Hanuman does it then nobody can take away the *cākara* ...

(Now Manoharan explains how people from another village may attempt to attract the *cākara* as well. They will do so by asking a different deity for help. These deities end up fighting each other over the *cākara*.)

Both deities (*devata*-s) fight each other to catch the *cākara*, just like in a battlefield. In a battle the one who is more powerful will succeed. The *nīcan* (impure one) wins always. The *nīcan* is more powerful than the others, the Brahmin's *pūja* always looses. The Brahmin's *karmaṃ* is truthful (*satyamulla*). But truth is meaningless (*ārtamiḷḷa*) nowadays. The truth will succeed only at the last. The same is the reason here: lies work first, the truth comes up later. The Brahmins do *pūja*-s in the way of truth. In the *pūja* they use the things [offerings] that are meant for the *satkarmaṃ*-s. They will not use palm-wine (*kallụ*) and chicken (*kōri*; i.e. non-vegetarian offerings). But in the *duśkarmaṃ*-s, chicken, goat, cow, buffalo may all be present. Most of the *devata*-s eat meat.

In this account, Manoharan mingles the role of mud in the formation of the *cākara* with the action of deities who fight over the place of its actualisation, as invoked by the rituals conducted by *karmi*-s or *mantravādi*-s, non-vegetarian sacrifices to little deities being those experts' primary mode of action. 'Sacrificial destruction implies giving something that is to be repaid', writes Marcel Mauss in his *Essay sur le don* (1954 [1925]: 14), and indeed the little deities seem forced by the *mantravādi*-s to move the *cākara* appropriately. On the other hand, it is clear that the deities are demanding, having to be worshipped and being brought offerings regularly.

While Manoharan describes Hindu ritualists, much of what he explains is mirrored by the doings of a renowned local Muslim expert. As a Thangal, this man belongs to a high-status lineage of Muslims who immigrated to Kerala from Yemen centuries ago. Like his Brahmin colleagues, he abstains from malevolent ritual action, which, like them, he considers to be only of short-lived benefit. Rather, he relies on his special relation (*pratyēkata bandhaṃ*) to the sea and to Allah. His father's father's father is reported to have invoked a *cākara* near Chamakkala about one century ago. Neither does this man follow a fixed sequence of ritual action nor does he draw on a fixed corpus of *mantraṃ*-s and *pūja*-s, but rather relies on what is revealed to him by Allah. Even though his conduct of rituals has varied over the years, it always includes the placement of *takeṭu*-s (small *yantraṃ*-s), copper plates with diagrams drawn and spells written on them. These are dug into the beach at the place where the *cākara* is meant to form. He does offerings of milk into the sea as well, explaining that these are meant to cool (inf. *taṇṇukkuka*) the sea and thus placate it. As such, he alludes to the concept of the sea's 'heat'.

Tinnan Nērca

Until the 1980s, the *tinnan nērca* (lit. 'betel vow', from inf. *tinnuka*, 'to chew betel'[15]), in short *nērca*, was a common ritual act performed in the context of fisheries in Chamakkala. Requested by the fishermen, it was performed by Muslims knowledgeable in prayer. When asking for it, the fishermen handed over betel leaves, areca nuts, lime and tobacco to the Muslim, who would then recite a series of prayers and Quranic texts in their name, including the *du'a* prayer and the Sura Yasin. The purpose of the *nērca* was to request the saint Mohiuddin Sheikh (Abd al-Kaadir) to placate the sea and make it flat. Such *nērca*-s had been performed mostly during the south-west monsoon when the sea was particularly rough. Ordered by individual fishermen, its effect was said to be collective, i.e. affecting a larger area of the sea. During twelve months of fieldwork in the village, I saw only one *nērca*, and this was probably performed to placate my persistent enquiries.

The abandonment of this ritual practice is commonly ascribed to the introduction of new technology and the old age of its practitioners, the most respected one being eighty-one years old at the time of my fieldwork. The decline in resort to this practice may reflect the disengagement of Muslims from fishing as well as the growing influence of a strict following of Quranic Islam disapproving of Muslim saint worship.

Modes of Relating

The material of Chapters 4 and 5 shows that Chamakkala fishermen speak about the sea with the vocabulary of human physiology, relate to it symbolically through ritual and perceive it in the context of their own position in Kerala's society at large, wherein they have been marginalised as fringe-dwellers. Perhaps most generally, the fishermen's activity in daily fishing and as patrons of rituals is to make inferences and act accordingly in an environment of intentional social beings, such as other people (including other teams of fishermen), deities, the sea and the *cākara*. The inferential schemes brought by fishermen to the sea and to the *cākara* are much like the ones they bring to social others. The fishermen find themselves in a social relationship with the sea and the task now is to grasp its essence anthropologically.

I take Marcel Mauss's (1954 [1925]) theory of gift exchange, understood by Alfred Gell (1998) as the prototypical anthropological theory of social relations, as my point of departure. In a famous passage of this work, Mauss (ibid.: 8–9) quotes Maori chief Tamati Ranaipiri's account of an exchange process between Maori priests, the *hau* of the forest and the hunters of birds, as it was recorded by Elsdon Best (Best 1909: 439). Mauss's text, in which he identified the *hau* as the 'spirit of the gift', has since fascinated many anthropologists, including Marshall Sahlins (1972), who noticed that Mauss omitted to mention that Ranaipiri actually described a Maori hunting rite, the ceremony of *whangai hau*. In his reinterpretation, Sahlins identifies the *hau* as the 'principle of increase' and criticises Mauss for trying to explain an economic concept in religious terms, whereas Ranaipiri actually expressed a ritual process in economic terms, that is, with the concept of *hau* (ibid.: 157; see also Godelier 1999).

When considering Ranaipiri's entire account of the *whangai hau*, the rite becomes clearer. In it the priests make offerings to *mauri* (fertility stones deposited in the forest) and the *hau* responds by providing the hunters with birds. The hunters must return part of their prey to the priests; only then is the priests' service compensated for. In yet another account of this passage, Gell (1998: 106–109) elaborates upon Sahlins's (1972) and

Babadzan's (1993) interpretations, noticing that the *mauri* and the birds are ritually one and the same thing, which the priests give to the hunters, intermediated by the forest's agency. The forest is obliged to give birds to the hunter, because it owes its fertility and productive capability to the priests' ritual action. In this account agency and fertility are potentially present in the forest, but have been co-opted by the priests in their ritual.

While the character of the *whangai hau* as a hunting rite makes it already clear that human-environment relations are alluded to in this passage of Mauss's essay, a few pages later its author draws attention to sacrifice as a mode of gift-giving that is meant to benefit from nature through offering to gods, for these 'gods who give and repay are there to give something great in exchange for something small' (Mauss 1954: 15).

This interpretation captures certain aspects of the ritual action of Chamakkala fishermen. Here it is the priest who gives *prasādaṃ* (the blessed flowers) to the fishermen, which they pass on to the sea for assuring its productivity. The fishermen pay the temple's priest from the money they have received from auctioning their catch. A cycle of exchange seems completed, and 'fish', so often called merely 'stuff' (*vastụ*) or 'substance' (*sādanaṃ*), are recognised as gifts in exchange. If so, the notion of killing, as well as the symbolic thought which a hunter's killing of game often involves, is bypassed. As elaborated by Valerio Valeri (2000) with respect to the Huaulu of Seram (Moluccas), killing game and the taboos associated with wild animals raise profound questions about what it means to be human. But in which sense do the fish caught in Chamakkala differ from the game that Valeri and other ethnographers of hunters have considered? Several differences suggest themselves. First, the fishes constituting the catch (typically oil sardines and mackerel) are small and numerous, not counted in number but in volume or weight of a catch (as measured by the standard *kottụ*, i.e. basket), thus inevitably perceived as lacking individuality. Secondly, individual fish are seen only after their death, unlike game that is perceived to interact visually with the hunter during the hunt (Ingold 2000), and thus the act of killing is anonymised. Thirdly, fish meat differs in its quality from that of mammals and birds, and perhaps this contributes to why, as Kurien (2000b) claims, 92 per cent of Kerala's population, including many Brahmins, eat fish.

A further observation supports the identification of fish as gifts. In trawling and purse-seining a part of the catch that is considered useless (perhaps because of a low market price) is commonly thrown back into the sea (where these fish are often bound to die). This is even done in the operation of Chamakkala *vaḷḷaṃ*-s and provokes the outrage of older fishermen. For them, to throw fish back into the sea meant to refuse a gift and to violate reciprocity. And to refuse a gift may prevent one from receiving further gifts in the future.

But altogether Maussian gift exchange remains insufficient in accounting for the fisherfolk's relation to the sea. In Chamakkala, there is no explicit local equivalent of the Maori *hau*, and unlike Ranaipiri's account of three partners involved in the exchange (priest, hunters and forests), here there are four partners (priest, fishermen, deities and the sea). As mentioned, it is with the involvement of a ritual expert that the ambiguous identification of *kaṭalamma* ('sea mother'; see Chapter 4) with the Hindu goddess or the sea is resolved.

The notion of *kaṭalamma* reminds one of the permanent asymmetry inscribed in the kin relation of a mother with her children; in terms of the flow of goods in exchange, it is mirrored by the sea's relationship to the fishermen. Nurit Bird-David (1990) draws on a similar observation in distinguishing hunter-gatherers, who conceive of their environment as 'parents' and are recipients of a 'giving environment' that provides them with food unconditionally, with those who consider their environment as ancestors (with whom, so she claims, exchanges demanding reciprocation take place). Neither scenario matches my data on the human-environment relations of Chamakkala fishermen sufficiently well.

It is apposite to start more generally and return to the direct, unmediated interaction between the fishermen and the sea. In wondering about it one is faced with the problem underlying all symbolisation. Roy Wagner (1977) has noted that, whether in biology, physics or anthropology, the more the units of a study are defined and thereby bound, the more difficult it becomes to study the relationships among the units. Conversely, the more the relationships are studied, the more difficult it becomes to define the units (ibid.: 386). In its logical structure, it is the anthropologist's version of Heisenberg's uncertainty principle of quantum physics, which states that in specifying the motion of a subatomic particle one can either determine its position precisely, at the cost of not knowing its velocity, or vice versa. A physicist has to decide whether to consider one or the other.

Much recent anthropological work on human-environment relations has focused on the units of relationship and has considered animals, plants, stones and other entities in the environment as persons. Examples are geese among the James Bay Cree (Scott 1989), animals and stars as seen by the Jivaro of Amazonia (Descola 1994: 93, 98), the sun as seen by the Ojibwa of Canada (Hallowell 1960: 36) and stones, rivers and trees as seen by the Chewong of Malayasia (Howell 1996: 128). What exactly is meant by 'personhood' in these instances and in what way the alleged 'personhood' of these entities may be comparable to that of humans is not always clear. This is an important deficit.

More recently, Bird-David (1999) and Tim Ingold (2000) have praised Irving Hallowell's (1960) inclusive, operational definition of 'personhood', which draws on Krech and Crutchfield's (1948) account of the perceptual environment (which they term 'psychological field'). To Hallowell, 'persons' are those entities to which a behaviour can be attributed that includes mobility, capriciousness and unpredictability, which are the perceived loci of causation, have 'power qualities' and are sensitive and reciprocally reactive, if they find themselves in morally sanctioned relations with other such entities (Hallowell 1960: 48). Illustrating this concept with his ethnography of the Ojibwa, Hallowell calls for the abandonment of human primacy in classification. However, since personhood, thus defined, is not an emic concept of the Ojibwa, one wonders if one has learned anything profound about Ojibwa culture at all. My material on the sea and the *cākara* matches all criteria of Hallowellian persons, but other than being perhaps of descriptive value all it tells me is that, in certain contexts, the sea and the *cākara* appear as agents, intentional social others.

Hallowell's concept of personhood lumps together elements that are distinguishable and may be independent of each other. One may consider capriciousness, causation and the embeddedness in morally sanctioned relations to inanimate entities as instances of anthropomorphism. However, as Pascal Boyer (1996: 89–90) reminds one, this is a broad category of projecting such aspects as anatomical structure, physiological processes, personal identity, social organisation and intentional psychology. Potentially, these projections arise from different cognitive processes. Psychological properties may be regarded as fundamental since they are 'almost invariably entailed by other projections, whilst the converse is not always the case' (ibid.: 90). This finding is affirmed by my material on the sea, in which aspects of morally driven responsiveness and intentionality dominate and physiological processes are evident, while its 'anatomy' is restricted to its 'standing up' and 'lying down'. Chamakkala fishermen are fully aware of the fact that speaking of the sea's physiology is reasonable only to a certain degree. 'The sea is mother' and the physiology of heat are counter-intuitive to them, much in the sense of Boyer (ibid.: 92), who finds that anthropomorphisms are counter-intuitive because of violating intuitive ontological expectations delivered by domain-specific principles, but in doing so they make use of intentional psychology, the 'richest domain of inferences available to human minds' (ibid.: 95).

The problem of symbolisation to which Wagner draws our attention, i.e. that one has to choose between studying units or relations, results in two complementary modes. Either one is 'relating the perceptibly differentiated, or differentiating the perceptibly relational, from the standpoint of the actor' (Wagner 1977: 391). Rather than dwelling on the unit of symbolisation, the 'sea as person', Chamakkala fishermen dwell on their relationship to it. Compared with the considerable number of relationships that constitute human persons (see Radcliffe-Brown 1952: 194), the sea is experienced in a more limited variety of situations and contexts. It may be conceived of with the kin category 'mother', but to the fisherfolk it remains a social other in a double sense: as non-self and as non-human. It is the (marine) environment, and in relating to the sea as environment the fisherfolk invoke what Wagner (1977: 398) calls a 'beneficient relational flow' between two partners that sustains them in fishing as the 'work of the west' (*paṭiññāttu pani*), and in their residence in the coastal area. The sea acts punitively in a rational way: it punishes evil and rewards good. Yet, as emphasised in the myth of the regular intimate meetings of the sea and the mountains (Chapter 1), the sea itself agrees to protect the people of Kerala in a contract-like agreement. This is implied when some fishermen say that the sea has no (traditional) right (*avakāśaṃ*) to come to the eastern land. With the coastal dwellers living in the sea's neighbourhood, the sea mirrors their human agency.

According to Wagner it is through conventions that a culture determines which semiotic mode is to be understood as the province of human action and intervention. Its complement, what is outside the reach of human action and intervention, will be conventionally regarded as the innate or as 'nature' (Wagner 1977: 393). In Chamakkala, invoking the 'relational flow' of truthfulness entails submission to a moral order that is associated with the sea as a rational and moral partner. The 'relational flow' between humans and the sea is limited by the sea's bodiliness as manifested by its physiology of 'heat'. This physiology cannot be affected from within the scope of a moral relationship. Unlike the moral relationship of mutual truthfulness, which is susceptible to human intervention, this physiology is perceived as innate, being outside human action and intervention.

Notes

1. The expression *satyamulla mutal kaṭal* is composed of *satyaṃ* (truth), the possessive marker *–ulla, mutal* ('from' [the beginning]) and *kaṭal* (sea).
2. Lat. *Ruta graviolens*, a small wild bush that is said to perish once touched by a menstruating woman.
3. Jack fruit tree (*Artocarpus heterophyllus*).
4. A mythical plant that is supposed to grow in the Western Ghats. All things touched by it are supposed to turn into gold (Shankunni 1998).
5. The Malayalam word *pula* is used for temporary pollution (birth, menstruation, death, impurity of work) as well as for a person's permanent pollution as the member of a specific caste (Warrier et al. 1999: 706).
6. In marked contrast particularly to Christians, Hindu women in Kerala are often restricted to the home or the compound. Osella and Osella see therein the articulation of a style of bourgeois housewife ethic (2000: 41–45). Some Araya women from Chamakkala commute daily for work, sometimes as far as the nearby towns of Thrissur and Chavvakadu. They are employed as shop vendors, as office staff in the panchayat office or as teachers. Visits

to the beach from Hindus living inland are usually restricted to funerals and *bali* offerings to ancestors at *karkaṭakaṃ* and *tulaṃ vāvu̥* and at Shivarathri.

7. These rules of pollution remind one of those applicable to sacred spaces. Writing on the performance of rituals in Sri Lanka, Obeyesekere notes that rules of purity and rituals of purification must be followed strictly to prepare the ritual arena to become a meeting place where gods, humans and demons can interact: 'If the gods are brought into contact with impure, particularly polluting, sources, the individual as well as the community will be struck with *vas* (ritual danger) – generally illness' (1984: 53). In this light it may be understood why some people say that the seashore will perish (*kaṭappuṟaṃ naśikku̥ pōkuṃ*) if approached by a (polluting) menstruating woman. Ram (1991: 79–85) has made similar observations among the Catholic Mukkuvar fisherfolk in Tamil Nadu.
8. Vishalakshmi (1973: 1006–1008) gives details on Araya funeral rites and pollution periods.
9. One lakh Rupees is 100,000 rupees. In 2000 this amounted to approximately 2,500 Euros. Such large amounts of money are at the most in the hands of well-off middlemen.
10. In the Bhagavad Gita, Parthasarathi is the conductor of Krishna's horse cart in the battle of Kurukshetra.
11. Vishnumaya Chattan is a non-Sanskritic deity that is popular among low-status castes in Central Kerala. According to myth, Vishnumaya is one of a great number (36, 39, 336 or 441) of Chattans that were born to the tribal woman Koolivaka after an apparently illicit relation with Shiva. Every Araya lineage temple in Chamakkala has at least one form of Chattan in its pantheon. Known beyond the borders of Kerala, Vishnumaya Chattan and some of his brothers are often linked to acts of sorcery (*durmantravādaṃ*; see Tarabout 1993, 1997a; Parpola 1997).
12. Since the mid-1990s there has been a ritual rivalry between the fisherfolk from the three beaches, two of them trying to top each other's offerings. This has resulted in three separate processions being held on the same day. In 2000, each one of these comprised between approximately 300 and 500 persons. Companykatavu's fisherfolk organised the most elaborate procession, involving an elephant carrying the *tali* from a temporary temple erected at the beach to the Palappetti temple.
13. Ritual experts do not belong only to the Kaniyan caste (traditional experts according to Thurston 1909: III, 178–200), but are found in all castes, including Brahmins and Pulayas (agricultural labourers). A number of Araya *mantravādi*-s work for the fishermen. The term *mantravādi* is applied to Muslim and Christian ritual experts as well.
14. This is a reference to the Ramayana epic, wherein the monkey-shaped deity Hanuman helps Rama to rescue his wife Sita from the demon king Ravana.
15. Besides these minor ritual acts, in northern Kerala the annual festivals of large mosques are called *nērca* (see Dale and Gangadhara Menon 1995).

6 State Interventions

A consideration of the fishermen's view of the marine environment and human–environment relations would be incomplete without considering the impact of the state, in the colonial period as well as since independence. In Chamakkala, its action has been directed at incorporating the fringe-dwelling population at the coast into its planning, development and regulation. In this chapter I focus on two state interventions in the fishery. The first is a ban on the fishery of oil sardines which was issued from 1943 to 1947 by the government of the (British-ruled) Madras Presidency. The second is a seasonal ban on trawling ordered by the state government of Kerala since 1988. I close this chapter with another example of state intervention, which, although not intended to affect fishing, has been understood by the fishermen as an unprecedented impact on the marine environment.

A Ban on the Fishery of Oil Sardines

In the early 1940s, an enormous decline in the landings of oil sardines was noticed in Chamakkala and elsewhere along the Malabar coast. The village had belonged to Malabar District, an administrative unit of the British-ruled Madras Presidency, since 1830. Old fishermen remember that for several years oil sardines were not caught at all. This was a serious issue. Locally, oil sardines are (and have been) the most important fish species for the nourishment of the population by virtue of its abundance and low price. Occasionally, this fish is called the 'provider of the family' (*kuṭuṃbaṃ pularti*). At the same time, fluctuations in the catches of this fish are well known. Within memory, sudden drops in the catches and, subsequently, years without any catch of oil sardines, had happened before. As such, it came as a surprise for the fishermen in Chamakkala when, in 1943, the Madras Department of Fisheries ordered a ban on the capture of small fry and spawners. The prohibitions were threefold and concerned

(a) the use of highly destructive boat seine (*mati kolli vala*) during the sardine season from August to April,
(b) the use of the encircling gill net (*mati cāla vala*) during the spawning period in August and September, and,
(c) the landing of more than a total weight of one maund [37 kilograms] of oil sardines under 15 centimetres length from any single boat during the fishing season. (ICAR 1971: 28)

Boat seines and encircling gill nets were the common gear to catch pelagic (i.e. near-surface) fish. The ban was renewed in 1945, when, in addition, the use of boat seines and encircling gill nets was forbidden all year round. Never before had a similar ban been imposed on the marine fisheries of Malabar.

When I learned about the ban during my fieldwork I wondered how old fishermen judged its usefulness. Did they think the ban benefited anyone? And, if so, whom?

Kunhiraman Kochikkattu, a man in his mid-eighties, remembers:

Kunhiraman Kochikkattu: In the sea no oil sardine was seen. At that time everybody was saying that the oil sardine was lost, then the police was watching if anybody was fishing oil sardine, they said that it is forbidden to catch oil sardines. But some [catches of] oil sardine arrived in Kochi and after some time small oil sardines got caught in some people's nets here, and because of this [illegal fishing] some police cases were filed. It is true, oil sardine was missing for some time, but it came back afterwards. That was when the British were ruling ... Sometimes fish is missing from the sea. Occasionally fish is not seen for one or two months and after that it comes here from other regions. The same fishes which left may come again here.

Chandran Kozhissery, a man in his late sixties, explains:

Chandran Kozhissery: At that time, oil sardine was missing in the sea for twelve years ... It may move through the sea and even to the eastern sea. That means to Madras. By moving like this the oil sardine vanished from here. There is only a small possibility that they turn around and come back here. Thus oil sardine became rare. Last year and the year before [i.e. around 1997/1998] oil sardine was missing a lot [as well]. The oil sardines which we got in the market were the oil sardine from Madras. To sell it here oil sardines were brought from Madras! Now they have turned to the northern sea, that means the Bay of Bengal, and then they reached the Arabian Sea. After breeding and hatching they have multiplied. Now they will be here for some time. They may move to the southern sea and then we will miss it for three or four years. Less many oil sardines were there for five years but in this year they have come back.

Given the age of this man, he cannot have witnessed the 1943–1947 ban himself, but has to rely on hearsay. His reference to recent decline in the fishery is appropriate. The Kerala-wide catches of oil sardine had plummeted from 190,000 tonnes in 1989 to a mere 1,554 tonnes in 1994, a decline of 99.2 per cent (Yohannan et al. 2000: 6). In Chamakkala, the oil sardine catches were thought to have returned to their former level only in 1999.

Ramakrishnan Cherupurayil, a man in his mid-seventies, recalls:

Ramakrishnan Cherupurayil: Oil sardine is the main food of shark. Previously the catch of oil sardines was banned by the government. It was not allowed to catch oil sardines, because oil sardines are the food of sharks. This was the reason according to the Fisheries Department. The order was issued that oil sardine must not be caught. Those who caught oil sardines had to be punished. Therefore, oil sardines were caught and secretly transported to far away places. The sale was also done secretly. If we were getting big sharks then we had to sell them to the Fisheries Department.

After half a century, the drop in the catches of oil sardines is an event that is well remembered by the fishermen of Chamakkala. The subsequent ban, the colonial government's attempts at its enforcement as well as some fishermen's attempts to circumvent it are equally well remembered.

Enforcing the ban locally was made the duty of the *kaṭakkoṭi*-s, local headmen of the Araya caste. At the time most of the *kaṭakkoṭi* families had not had active fishermen for generations. Many of them had turned into petty landlords, holders of small government posts and teachers in the state 'Fisheries schools'. As such, the fishermen experienced the complex formation of a 'state' through their interaction with individuals to whom they might entertain a variety of social relationships. The distinction between 'the state' and 'the local population' was blurred. At least for the time of the ban, the boundary of the colonial state, diffuse as it may have been, was situated within the fishing community itself (cf. Gupta 1995).

To sum up, older fishermen insist that the fluctuations in sardine fishery were not due to any human intervention but reflected 'natural' processes beyond human influence, and none of the fishermen believed that the ban contributed to the recovery of the catches after 1948. Rather, the ban was judged as a misconception and, as such, a meaningless endeavour. At independence in 1947 the ban was abolished because of 'difficulties encountered in its enforcement, the lack of preventive staff over a long coast line and of similar legislation in adjacent states' (ICAR 1971: 28).

The Madras Department of Fisheries

Why, then, did the Department of Fisheries consider the ban necessary? The fishermen could not know that it resulted from the first scientific investigations into the marine fisheries of South India. After all, during the ban (as well as before) visits of British officials to the seashore were rare and did not involve much interaction with the fishermen. None of the old fishermen of Chamakkala can remember these officials discussing the reasons for the installation of the ban with them.

In order to shed light on why the fisheries department installed the ban one must consider its intentions to develop the fisheries in more detail. From its foundation in 1908 the Madras Department of Fisheries considered the fisheries an issue that was to be developed according to Western technical and scientific aspects.[1] In that year Frederick Nicholson became the first Director of Fisheries in Madras. His verdict on the fisheries in Malabar as of 1916 is worth quoting:

> [The] fishing industry is in the most primitive condition, quite undeveloped in any of the modern methods and allied industries, bound by custom and ignorance, and entirely without initiative in new departures; it is the Government officers only who have a larger knowledge and a certain degree of initiative, and it is, at present, for them to lead the industry and the men, as has been done in the oil and guano development, in canning and other curing and cultural methods, and as will be done shortly in matters of capture; this is the raison d'être, and this only, of the Government Department [of Fisheries]. (Govindan 1916: 2)

Thus, the artisanal fisheries of the Malabar coast were not seen as what they were, but as what they were not – and as what they should be. The British intervention was based on a paradigm that was rather different from the fishermen's idea of interacting socially with their environment. Nicholson believed fishing to be an industry that had to be developed by the government. His comparison of fishing with the developments in the oil and guano productions and the hint at canning technologies suggests that he intended to develop Malabar fisheries in an export-oriented way, rather than merely wishing to improve the living conditions of the fisherfolk. This development was to be based on 'fisheries science', then newly emerging in Europe. Of course, 'science' was understood as the production of knowledge according to the methodological style inherited from Western modernity, implicitly representing, at any time, the closest possible approximation to 'truth'. If artisanal fishing was indeed regarded as 'bound by custom and ignorance', one may safely assume that the British had little interest in consulting fishermen about their ideas on the fluctuations in the landings of oil sardines. Indeed, in all the extensive reports published by the British administration hints of any interest in the fishermen's knowledge are scarce. It was James Hornell, who became Nicholson's successor as Madras Director of Fisheries in 1918, who occasionally noticed the fishermen's ideas in his publications (see Hornell 1910: 97, 1931: 54), albeit without allowing it to influence research schemes or government policies. This is not surprising, since such an interest would have run counter to the department's intentions. By criticising the 'ignorance' of the Malabar fisheries of the day and their 'being bounded by old customs', Nicholson questions their authority. In contrast he notices that it was only Government officers who had a 'larger knowledge' and a 'certain degree of initiative', thus legitimising the doings of the colonial administration. 'Science' and 'scientific knowledge' were of strategic significance in legitimating British colonial intervention in India (see Kumar 1995, Prakash 1999).

While the politics of the Madras Department of Fisheries must be seen in the context of the British colonial project, its initiatives had a long-lasting effect on the fisheries of Kerala. Besides initiating a chain of Fisheries Schools along the coast, Nicholson initiated improved health facilities. In 1918, his successor James Hornell started a programme of fisheries cooperative societies. The latter were meant to improve the fishermen's control over the sales of fish, to provide credits and to initiate shared processing facilities, especially fish curing-yards (Kurien 1980: 1; Udayabhanu 1990: 125–128). A similar development took place in Travancore where the first fishermen's cooperatives were formed in 1919 (Udayabhanu 1990: 143). By 1933, there were ninety-five cooperatives with 8,194 members (Kurien 1985: A71). Nevertheless, these cooperatives were not considered a success. When a state committee had to evaluate the status of the cooperative societies in Travancore and Cochin, it found them in a dismal condition. Lacking the support of community leaders, they were sabotaged systematically by the middlemen (ibid.). The condition of the cooperative societies in Malabar was probably not much different.

A Scientific Intervention

Even before Nicholson took up his position, the fish resources of the Madras Presidency had been estimated in 1899. In a note on the 'desirability of developing the agricultural department [of Madras Presidency]' it was estimated that one million tonnes of fish could be fished annually (Anon. 1915: 2). This had been the first estimate against which the further development of the catches was to be measured.[2] Previously, only exports of fish products (mostly dried fish and fish oil) had been recorded. Now the aim was to quantify fish landings in a comprehensive way in order to understand the economic potential of the fishery. Hitherto, the catches of the artisanal fisherfolk, which were mostly consumed locally, had remained unknown. In 1908, the Department of Fisheries began to record the catches of artisanal fishermen and to publish them in its annual reports. Quantifying resources in such a way is a typical activity of a modern state government (see Scott 1998).

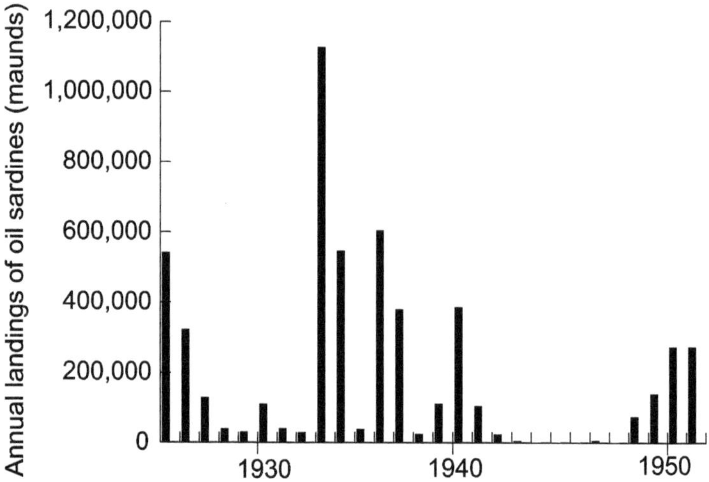

Figure 6.1: Annual landings of oil sardines in Malabar District (in maunds; 1 maund = 37 kilograms). Source of data: Nair (1952).

Soon the numbers in the landings revealed immense fluctuations. Of all the fishes representing the catch, the fluctuations in the fishery of oil sardines were the largest. This was a matter of concern, since the British had recognised that this fish was important for the nourishment of the local population. In Malabar District, the catches of oil sardine were low in 1908–09 to 1911–12, reasonable in 1912–13 to 1913–14, then dropped again from 1914–15 to 1918–19 (Balan 1984: 1). Throughout the 1920s and early 1930s the catches fluctuated widely between 1,100 tonnes (in 1931–32) and 41,690 tonnes (in 1933-34; see Figure 6.1). In the Department of Fisheries' *Annual Report* for 1941, we read: 'So great are the fluctuations [in the catches of oil sardine] that they render the fishery unstable and undependable' (p. 7). Elsewhere, the fluctuations were called 'unfortunate and disastrous' (Devanesan 1943: 2).

When James Hornell was appointed as Director of Fisheries in 1918, he initiated scientific research into the nature of these fluctuations. The greatest step towards this end was, in 1921, the inauguration of a fisheries research station on West Hill in Calicut (today's Kozhikodu): 'It is only an institution which gives facilities for laboratory work, study of collections, and for consultation of books, in a word, a Fisheries Research Station that can with a definite plan of work promote, direct and consolidate marine fisheries research' (Devanesan 1943: 1).

Studies in the fluctuations of the oil sardine fishery began at the station in 1921 itself with the attempt to learn more about the life cycle of this fish and its spawning grounds. The results were published three years later (Hornell and Nayadu 1924), but were largely inconclusive. Following a decade of immense fluctuations in the catch, an improved investigation was begun in 1930, when D.W. Devanesan was appointed as the Assistant Director of Fisheries (Biology). A biologist with a Ph.D. from London, Devanesan was to implement a programme of state-of-the-art research at the Calicut station. His specific duty was to address the issue of the fluctuations in the fishery of oil sardines. The advances of his research can be traced in the annual reports of the fisheries department until 1942 and in a separate report (Devanesan 1943). Unfortunately, no annual reports were published during the years of the ban between 1943 and 1947.

I have mentioned that, from the viewpoint of local fishermen, the border between the 'colonial state' and their caste might have been blurred due to the enforcement of state politics by fellow caste members. On the other hand, the Department of Fisheries did not perceive an absolute border between the 'colonial government' and the local population. This may be surmised if one notices that Devanesan and other Indian co-workers of the department appear as authors and co-authors of its publications. The title page of Devanesan's 1943 report on his investigations of the fishery of the oil sardine mentions not only the Ph.D. he had obtained from the University of London, but calls him 'Rao Bahadur', a title given to Indian officers, and addresses him respectfully as 'Sahib'.

The career of some other co-workers in the department is equally remarkable. Particularly noteworthy is V.V. Govindan, an Araya man from Kozhikodu, who was hired by Nicholson in 1914 and soon promoted to become the 'Assistant Director of Fisheries (West Coast)'. By hiring local men the department must have attempted to improve the realisation of its aims of development. Prior to his promotion as Assistant Director, Govindan was sent on a journey to Britain and Germany in order to study what Nicholson had called the 'modern methods and allied industries' of fishing. As Nayadu and Devanesan did some years later, in 1916 Govindan published a report in the *Madras Fisheries Bulletin* under his own name. Until today, Govindan's career is remembered with much respect in Chamakkala, especially among academically educated Arayas. This was made particularly clear when, in 1978, the *Rao Bahadur V.V. Govindan Memorial Souvenir* was published, a 200–page book containing articles in praise of Govindan's life and work. It was published on the occasion of his 100th birthday in Kaipamangalam, the village south of Chamakkala.

While the recordings of the annual catches of all fish species continued throughout, Devanesan initiated size measurements of large numbers of oil sardines. In doing so, his aim was to investigate whether or not the fluctuations were due to overfishing and if so, how quickly the stock recovered. If years of large catches were indeed followed by years of small catches, overfishing was likely. Besides overfishing, there were two other hypotheses. First, the 'migration theory': that oil sardines migrated away temporarily (as the artisanal fishermen surmised) and, secondly, that shoals of sardines were prevented from entering the coast due to specific hydrographic conditions (rainfall, temperature, salinity, etc.). The basic question underlying all of these ideas led back to the issue of overfishing. In 1933–34, Devanesan formulated it as 'unless the Malabar sardine proves to be a distinct local race confined to that coast, the question of depletion by fishing does not arise, though this has often been alleged to be the reason for the scarcity of fish in some years' (cited in the *Annual Report* for 1941, p. 7).

Indeed, if the Malabar oil sardine is a local fish, then its fishery could be influenced by a ban on the fishery. On the other hand, if it is not local, then a regulation that is enforced locally must have only limited success. This was the 'local race theory'. If verified, it implied that one could consider the abundance of oil sardines in terms of a 'stock', quantifiable and, thus, in principle, manageable by government regulations.

Devanesan surmised that whether or not the Malabar oil sardine is a local fish could be found out in two ways. First, its spawning grounds must be located somewhere in the Malabar area. And, secondly, there should be anatomical differences in oil sardines from other parts of the Arabian Sea and Indian Ocean. Results on the first issue were published in the *Annual Report* for 1935. It reported the discovery of spawning grounds in the sea about fifteen kilometres off Calicut. Resolving the second issue took more time. Oil sardine specimens were collected from Karwar (near Bombay), Karachi, Muscat (Oman) and, later, Aden (Yemen). Some additional specimens were consulted in the collections of the Natural History Museum (London). The differences in anatomy were slight and thus an unambiguous conclusion seemed impossible. Considering the development of landings and size measurements in the last decade seemed more promising. It revealed that large catches of small, i.e. juvenile, oil sardines (with sizes under fifteen centimetres) in one year were followed by a lack of small oil sardines in the next year. Apparently, the fishery of juveniles had prevented the stock's replenishment. Devanesan concluded that the 'migration theory' was unlikely and that the 'local race theory' might have to be preferred (1943: 36). In other words, he surmised that overfishing of a local 'stock' of oil sardines was occurring. With this claim made, government regulations could be argued for and the state could take the management of marine resource under its auspices.

It was only at this point that Devanesan and his colleagues consulted the artisanal fishermen. Wondering about possible consequences of this insight, the researchers of the Calicut station went to the local fisherfolk, trying to find out with which net small sardines were caught. It turned out to be the small-meshed boat seine net (*mati kolli vala*). When the scarcity of oil sardines noticed in 1940 and 1941 continued in 1942, usage of this net was to be banned for nine months per year starting in 1943. As mentioned above, the ban was renewed and intensified in 1945 before being abolished in 1947. The landings of oil sardines recovered shortly thereafter (see Nair 1952; Balan 1984: 1). From the fishermen's viewpoint, the ban was the misconceived attempt to influence an entity (the 'fish stock') that did not exist.

Banning Monsoon Trawling

Half a century later, the state's intervention to regulate marine resources has been demanded by fishermen's organisations from south Kerala. In June 1999, a heated public debate flared up in Thiruvananthapuram, the capital of Kerala since its foundation as a unit state in 1956. Participants of a demonstration organised by the Kerala Swathanthra Matsyathozhilali Federation (KSMTF; a Christian-dominated fishermen's union) gathered in front of the Secretariat, the seat of Kerala's Chief Minister, and demanded the installation of a three-month ban on trawling and purse-seining during the south-west monsoon, plus legislation preventing deep sea fishing by foreign ships in India's Exclusive Economic Zone (EEZ, extending about 370 kilometres into the sea). In trawling a bag-shaped net is dragged over the sea bottom by a 'mechanised' boat, a deep-keeled vessel with an inbuilt engine. Purse-seining resembles the operation of ring seines as described in Chapter 3, except that the nets are much bigger and the boats in use are mechanised as well, unlike 'motorised' boats such as the plank-built *vaḷḷaṃ*-s in use by Chamakkala fishermen, to which outboard motors are merely attached. Mechanised boats cannot be operated from a beach, but only from harbours. The distinction between the 'mechanised sector' of trawlers and purse-seiners, a capitalist mode of fishing involving owners of equipment and labourers who often do not belong to any of Kerala's fishing communities and aiming at the export of prawns for foreign markets, and the 'artisanal sector', allegedly dominated by 'traditional' fishermen working for their subsistence, has run through political debates about Kerala's fisheries since the 1970s.

The debate on foreign deep-sea fishing vessels entering Indian water commenced in 1991 when India conceded to the World Bank to 'liberalise' its economy and let foreign investors in. The months June, July and August mark the principal spawning period of many economically important species (including oil sardines, mackerel, prawns, anchovies, etc.). The ban was meant to allow these juvenile species to grow and the fish stocks to recover. A three-month ban on trawling and on purse-seining during this period was recommended by several expert committees of fisheries scientists, yet only after massive protests from south Kerala's Christian fishermen did the state government first install a trawling ban in 1988. By the mid-1990s, the trawling ban was regularly installed each year for forty-five days (from mid-June to the end of July). Purse-seining was banned annually for two months (July and August). This is much less than demanded by artisanal fishermen and deemed useful by the fisheries scientists, but too much for the owners of trawling boats, who reacted by counter-protests and intense lobbying. The campaigning of both camps has developed into an annual political ritual, as the columnist Anand Parthasarathy poignantly expresses: 'We have seen the same pathetic farce played out every year: a ban imposed half heartedly for six weeks or so, when the country fisherpeople begun to agitate, and then prematurely lifted when the mechanised lobby exerts its own political pressures' (*The Hindu*, Kerala edition, 7 June 1999).

This laconic commentary notwithstanding, the fishermen's agitation has been deemed widely, and internationally, a success. Just a week after Parthasarathy's commentary was published, Thomas Kocherry, a Catholic priest and renowned leader of the fishermen's movement in southern Kerala, was awarded the Sophie Prize in Oslo, Norway. Established by the Norwegian author Jostein Gaardner and his wife Siri Dannevig, this award rewards outstanding persons promoting environmental protection and sustainable development.[3]

Three hundred kilometres further north, on the beach in Chamakkala, there was little excitement about the debate going on in Thiruvananthapuram. Even though trawlers have been a daily sight here (on some days I have counted more than one hundred trawlers; Figure 6.2), fishermen calmly discussed the newspaper reports about what happened in the state capital. This does not mean that they were not concerned about the impact of trawling. In the 1980s, fishermen of Chamakkala had participated in statewide protest against trawling, but its outcome was regarded as ambivalent. They note with scepticism that there are only five

Figure 6.2: A trawler operating within five kilometres of the coast.

fisheries stations to control the enactment of the ban along a coastline of more than five hundred kilometres. Officially, the Chamakkala area is to be controlled by the station at Vypeen near Kochi. It is equipped with just one fisheries boat.[4] None of the fishermen could believe seriously that this boat is able to control the movements of hundreds of trawlers which are based at the harbours in Kochi, Azhikodu and Chavakkadu. Oftentimes, trawlers are seen fishing within ten kilometres of the coast, a violation of the Kerala Marine Fisheries Resource Act (KMFRA), installed in 1980 in order to protect artisanal fishermen (Government of Kerala 1983). Filing cases against such trawlers at the police or at Kerala's High Court rarely results in an indictment of the culprits but means getting involved in the scorned machinery of bureaucracy. And more than that, the idea of a fish 'stock' and the effects of the trawling boats are seen in a very different light from that of public discourse, as I shall elaborate further below.

For many years, Kerala's post-war fisheries development seemed like a story of success. The total catch landed in the state rose steadily from 152,000 tonnes in 1956 to 448,000 tonnes in 1973 (Kurien 1985: A79; Kurien and Thankappan Achari 1988: 2013). In 1971–1975 the average catch amounted to 406,000 tonnes per year, but this declined to 332,000 tonnes in 1976–1980. The contribution of the artisanal sector fell by 50 per cent from 335,000 tonnes in 1969–1971 to 167,000 tonnes in 1980–1982. In this period, the share of the artisanal sector in the total catch dropped from 89 per cent to 57 per cent (Meynen 1989: 745). Each artisanal fisherman had caught on average 3,340 kilograms of fish per year in 1969–1970, but only 1,780 kilograms in 1979-1980. The annual availability of fish per capita declined from 15.2 kilograms in 1956-58 to 9.2 kilograms in 1982–83 (Kurien 1985: A79). The share of the artisanal sector in Kerala's prawn fisheries declined from 95 per cent in 1962 to 24 per cent in 1985 (George 1988: 2). These statistics illustrate the development of Kerala's fisheries which began to emerge in the 1970s. First, the mechanised sector, which had grown since the early 1960s, had developed into serious competition or even a threat to the artisanal sector. Secondly, the fisheries resources became over-exploited.

Statements that overfishing did happen in Kerala have been made since the late 1970s by researchers associated with the Fisheries Research Cell of the Programme for Community Organisation (PCO), a Thiruvananthapuram-based non-governmental organisation of the (Catholic) fisherfolk. The experts included a retired joint director of fisheries of the government-run Central Marine Fisheries Research Institute (CMFRI) (George 1988). Their studies pointed out that overfishing was twofold: first, increased fishing effort did not result in more financial gain (economic overfishing) and, secondly, the catch dropped despite increased fishing efforts (ecological overfishing). This hinted at a decline of the fish and prawn resources.

In South Kerala, where the density of mechanised boats was highest, the artisanal fisherfolk reacted with sustained political agitation and demanded motorised craft and improved gear for themselves in order to be able to compete with the mechanised boats. The political agitation was aimed at the proclamation of an exclusive zone for small-scale fishermen. The ban of destructive fishing techniques was meant to restrict trawling and purse-seining as two capital-intensive technologies inaccessible to artisanal fishermen.

The Christian fisheries activists came up with two arguments against trawlers and purse-seiners. First, attention was drawn to scientific studies that seemed to prove that overfishing did indeed happen. While the fisheries activists ignored local environmental knowledge, they argued, secondly, that artisanal fishermen are especially connected with the sea. Emphasis was put on the concept of the sea as *kaṭalamma*. The fisherfolk would then be the 'children of the sea', a notion much popularised in publications, leaflets and posters. It is not obvious whether practising fishermen would agree with this term. As Cecilia Busby (2003) has shown in her insightful study, this notion transferred the claim of rights of access to marine resources and the protest against a capitalist mode of fisheries to a claim of identity, i. e. the artisanal fishermen's exclusive kin-like connectedness with the sea versus the non-connectedness of boat owners and workers.

Arguing with scientific studies and claiming an exclusive connectedness to the sea differs much from what Chamakkala fishermen would find convincing. As shown above, they do not share the notion of a fish stock, essential to all argumentation about banning a fishing technique unequivocally. And, while sharing the concept of *kaṭalamma*, they do not consider themselves the exclusive children of the sea. Unlike the claim to an exclusive kin-like connectedness with the sea that is lacked by other social groups, Chamakkala's fishermen insist on the open access nature of the sea and emphasise the performative character of truthful conduct (Chapter 5). This attitude was clearly expressed with regard to defining access to the local *cākara*, which – unlike several other *cākara*-s of Kerala – is open to anybody who wants to fish there.[5]

In south Kerala in the 1980s, the fisherfolk's agitation was initially led by Catholic priests and was largely organised by the Christian-based KSMTF (Meynen 1989). Activities included demonstrations, road and railway blockings, sit-ins and hunger strikes. These attained a broad coverage in the media and imposed pressure on the government. The boat owners emerged as an enemy and began to organise themselves in a union, emphasising that, by bowing to the demands of the artisanal fishermen, the government would risk the loss of foreign earnings and employment.

Nevertheless, the government reacted with the implementation of the KMFRA in 1980. A committee was installed to evaluate the necessity and potential benefit of a ban on trawling and purse-seining during the monsoon. This committee, termed the Babu Paul commission after its convener, presented its final report in 1982, but did not arrive at an unequivocal recommendation. Another committee, the Kalawar commission, was formed in 1984 and delivered its report in 1985. This commission did not demand a monsoon ban of trawling and purse-seining either, but recommended a strict upper limit for the number of such boats in Kerala. There was no immediate political reaction to the commission report, even though already in 1973–74 the number of registered boats (2,036) was twice the maximum limit as recommended by the commission (1,045). In 1980 the number of trawlers in the state was estimated to be about 3,000 (PCO/SIFFS 1991b: 4).

In the late 1980s, the Kalawar commission was followed by the Balakrishnan Nair commission, which, like its predecessors, was composed of fisheries scientists of the Central Marine Fisheries Research Institute (CMFRI) in Kochi and high-ranking officials of Kerala's fisheries department. Its report, released in June 1991, is full of statistics and graphs, underlining its claim to be a scientific document (Balakrishnan Nair 1991).[6] Since then, a number of further reports have been produced, typically being discussions of the development in the landings rather than actual biological studies of the fish populations in the Arabian Sea. In a sense it was the fisheries activists who challenged the government to respond scientifically and demanded its action to be legitimised by knowledge which, at least in its epistemological foundations, is extraneously produced.

Before considering how Chamakkala fishermen judge the influence of mechanised boats and motorised boats on the sea and on fishing, we need to know more about how they came to their large number.

A Brief History of Trawling and Motorisation in Kerala's Fisheries

At India's independence in 1947, the National Planning Committee judged the fisheries of Kerala to be 'backward and primitive' (Kurien 1985: A72), a verdict not much different from that made by Frederick Nicholson thirty years earlier. No attempt had yet been made to introduce new fishing techniques, contrary to Nicholson's promise of 1916. Shortly after the formation of Kerala as a union state, a coalition of communist parties was elected to power in 1957. It declared the improvement of fishing as a source of livelihood for the fishermen and as a source of food for the population as the aim of fisheries development. Fishermen were to be organised by strengthening the cooperative societies. Introducing nylon nets was meant to improve their productivity.

At this time a major impact on artisanal fisheries in Kerala was under way. In November 1952, the governments of India and Norway had signed an agreement for a fisheries development project. This Indo-Norwegian Pilot Programme (INP) was financed by the Norwegian government and led by Norwegian fisheries experts (Klausen 1968; Platteau 1984; Kurien 1985; Meynen 1989). It was the first development programme under the auspices of the United Nations. While the Indian government had not requested such a programme, it agreed under the condition that the project was aimed at providing food for the rural population, but not for export. This was to be achieved through providing active fishermen with more efficient technology (Platteau 1984: 80; Kurien 1985: A73).

The programme was begun in 1953 in the villages Sakthikulangara and Neendakara in south Kerala's Kollam District. First steps were taken in fitting outboard motors to existing craft. When these were found unsuited for such a purpose, experiments with new boat designs were undertaken, but none of these proved superior to the ones used by the artisanal fishermen. While for INP's Norwegian experts the pressure grew to produce a success, a private merchant independently decided to arrange for the export of thirteen tons of frozen prawns to the United States. These had been caught by artisanal fishermen from the Kollam area. The inshore waters of Kerala had long been known to have a rich resource in prawns, but remained largely untapped because of the low demand for prawns at local markets. This export deal proved to be an enormous success. In 1957, six merchants exported a total of 458 tonnes of frozen prawns to the US (Kurien 1985: A74). Soon prawns were exported to Japan as well and by 1962 the export amounted to more than 2,000 tonnes and continued to rise thereafter. In 1961, the sale of one tonne of frozen prawns yielded 4,000 rupees after subtraction of packing and shipping costs, compared with a value of 150 rupees for fresh fish sold locally (ibid.: A75–A76). Prawns, which had been used as manure for coconuts previously and had been cheaper than mackerel, turned into 'pink gold' (Kurien and Thankappan Achari 1988: 2012).

For the INP experts, the prawn export boom suggested a new direction. In 1957 freezing technologies were introduced to the project area. These were intended to make possible the supply of frozen oil sardines to markets away from the coast. But the costs were too high to make the sales economically viable. The INP experts decided to concentrate their efforts on prawn fisheries and to make trial exports of frozen prawns to the US. This was clearly against the project's initial outline and the decision was made against the initial resentment of Indian representatives on the INP board. The freezing facilities were well suited for the prawn export business. In order to make the catch of prawns more efficient, bottom trawling was introduced in 1960 (Klausen 1968: 123). By introducing trawling to India, the INP experts imported a fishing technique that had never been permitted in Norway's inshore waters due to concerns about its harmful effects on fish stocks (Kurien 1985: A76).

When the INP was closed in 1963 it was evaluated as a success, even though virtually all of its initial conditions had been violated: industrial techniques were employed and the prawns, which were caught for export, bypassed local consumers. Mechanised boats capable of trawling were given to fishermen's cooperative societies that had been founded for this purpose. The profit made by these societies was essential in order to consider the INP a success. But soon these societies were dominated by middlemen and moneylenders, for whom the INP provided the expertise of developing a profitable export-oriented prawn fishery. When the programme ended, the facilities for freezing and processing were leased to merchants, who introduced more trawling boats and began to invest in fishing on a larger scale. Workers were hired to operate the trawlers. A capitalist mode of fisheries began.

The Norwegian intervention continued after 1963 with the establishment of 'integrated fisheries projects' in Kochi, Karwar (Karnataka) and Mandapam (Tamil Nadu). These were meant to develop harbour facilities for prawn exports. By now in need of foreign currency, the Indian government welcomed these projects (Kurien 1985: A76). In order to reach all the fishermen of the state, a new programme was commenced to supply groups of artisanal fishermen with mechanised boats and the facilities for bottom trawling. To be eligible for such a loan, small cooperative societies, the *Matsya Ulpādaka Sanakaṭṭe Saṃkhaṃ*-s (MUCS; 'fish production societies'), were founded. Groups of active fishermen could jointly apply for a highly subsidised mechanical boat, called a 'society boat', and for loans to make its operation possible. Due to this programme the number of fishermen's cooperative societies in Kerala grew from 241 in 1958–1959 to 1,036 in 1973–1974. Until the mid-1970s, 805 mechanised boats capable of trawling had been issued to artisanal fishermen. In 1977–1978 the programme was dissolved and the societies were liquidated. It had turned out that the societies did not benefit the active fishermen, but had been taken over by a minority of rich and influential men (Kurien 1980: 2–3). With the introduction of the 'society boats' and the increasing number of privately owned trawlers, the number of mechanised boats in Kerala increased from 115 in 1958–1959 to 2,105 in 1973–1974 (ibid.). Like other mechanised boats, the 'society boats', being deep-keeled and heavy, had to be operated from a harbour. This meant that Kerala's coastline came to be dotted with harbour facilities for trawlers and other mechanised boats. The society boat that was run by Chamakkala fishermen was operated from Azhikodu, twenty kilometres to the south.

In the mid-1960s, trawling boats came to be a common sight from the beach in Chamakkala. Local artisanal fishermen began to consider this presence as an encroachment on their fishing territories. The priority of artisanal fishermen for access to marine resources had not been debated before. Occasionally, trawling boats damaged the equipment of artisanal fishermen (especially gill nets) by running over them with their boats. When this once happened in 1974, a serious conflict emerged. The trawling boat that had done so was surrounded by *vañci*-s and was captured by local fishermen. A fight began and more and more men entered the scene. One Araya man, who had worked on the trawling boat, was killed. Several men were arrested but ultimately set free for lack of evidence. Bottom trawling placed members of the fishing community on two sides of a divide, diluting the border between 'good' and 'bad' that could be perceived initially.

Matsyafed

Attempts to motorise artisanal fishing craft had been made since the 1950s (PCO/SIFFS 1991b: 5). There was no general impression that the use of these motors was economical and most of these projects were quickly abandoned (Kurien and Willmann 1982). The growing fraction of mechanised boats on the fish landings reinforced the impression that it was unavoidable to equip beach-landing artisanal craft with motors as well.

The state government supported the motorisation of craft with subsidised loans. These were issued through the Department of Fisheries. It was hoped that this would weaken the artisanal fisherfolk's feeling of being disadvantaged. To the Congress-Party-led government, elected in 1982, the politics regarding this issue was aimed at a delicate balance between the demands of the fisherfolk, who represent a significant 'vote bank' and were influential through the media coverage of their agitation, and the boat owners with their strong political influence. At the same time, for the artisanal fishermen to argue for the motorisation of the craft implied undermining the ecological issue of overfishing. The introduction of ring-seine nets by artisanal fishermen in the mid-1980s and the installation of motorised winches to pull large nets onto the boats has made their claim to sustainable fishing even more doubtful. In southern Kerala a new plywood boat design was introduced by the Intermediate Technology Development Group, a UK-based non-governmental organisation, in central and northern Kerala attempts were made to fit small (8 hp) outboard motors to existing *vañci*-s. The Department of Fisheries gave loans to individual fishermen only.

Fishermen cooperative societies had been defunct since the 1970s. Just one of the MUCS cooperatives was still operating. When the KMFRA was passed in 1980, the Kerala Fishermen Welfare Societies Act (KFWSA) had been passed along with it and 222 village-level cooperatives were formed (one for each beach, *kaṭavu*). In 1984 Matsyafed, a government-controlled apex society, was founded. Henceforth, the majority of government loan and support schemes, as well as loan schemes issued through the central government (such as loans of the National Co-operative Development Corporation, NCDC) have been issued through Matsyafed. In addition, there are insurance schemes for boats and motors, the subsidised motors and gear are sold through a chain of stores and educational programmes are supported. As it was the case for the cooperative societies that had been founded at the beginning of the century, the official primary objective was to weaken the influence of the middlemen (Matsyafed 1999: 1).

Most fishermen consider the schemes run by Matsyafed successful. The rapid spread of motorisation is often associated with the foundation of Matsyafed and with the efficiency of its schemes. The number of motorised artisanal boats grew from about 50 in 1978–79 to about 9,000 in 1988 and about 17,000 in 1995.[7] Upon their introduction it was decided to use kerosene as a fuel since heavily subsidised kerosene is issued for household purposes via the government-run ration stores.[8]

In Chamakkala, Matsyafed loans are issued through the local cooperative society, based at its office at Four Cross junction. In 1999–2000, it had 1,600 members from Chamakkala and its two adjacent villages, supplying loans to eleven *vaḷḷaṃ* units and fourteen *muṭuveṭṭi*-s. Since 1996, all teams that receive loans from Matsyafed have their catch auctioned by a Matsyafed-appointed auctioneer. One of Matsyafed's sixty-one Cluster Project Offices is situated next to the cooperative's office in Chamakkala.

A Sea Made Restless

While Chamakkala fishermen have participated in the statewide trend of motorisation, partly aided by Matsyafed, their view of the impacts of trawlers and the reasoning for demanding the implementation of a ban differ much from those of south Kerala's fisheries activists.

Despite some early protest against trawlers, Chamakkala fishermen consider trawling to have an ambivalent effect on local fisheries. During the second fishing trip described in Chapter 3, I was astonished by the equanimity of the men on the Ponnonam *vaḷḷaṃ* when we were surrounded by almost one hundred trawlers in the sea. I anticipated that these must have been considered a problem. Abdu, on that trip the *aryakkāran* (skipper), explained to me that the trawlers would not generally be considered in such a way:

Abdu Puzhamkaraillathu: [The trawlers] catch the fish that we won't get. Only during the monsoon will we feel [i.e. be affected by] the trawling. But even then there is no problem. If we demand them to go away they will do so.

While *vaḷḷaṃ*-s are aiming to catch near-surface (pelagic) fish, trawlers are believed to catch solely the bottom-dwelling (demersal) fish. This may give the impression that *vaḷḷaṃ*-s and trawlers are aiming at mutually exclusive fish species and do not compete against each other. But in fact the artisanal fishermen know well that trawling affects the marine environment more profoundly. Abdu admits that the noise of the trawlers' engines scares the fish. Knowing this, he tried to make use of this effect in advancing to the front of the trawlers, anticipating (in vain) that the fish would be chased there.

Nevertheless, trawlers are considered to produce negative effects as well, for fish may be assumed to be scared not only in one particular direction. Shoals may also be dispersed, with the fish moving to greater depths in the sea, where they cannot be caught by artisanal fishermen. If the trawlers operate in the outer sea (*puṟaṃ kaṭal*) far away from the coast, they are believed to benefit the artisanal fishermen by driving the fish towards the coast. Joseph explains:

Joseph Parayattu: Trawling is not always a *dōśaṃ* (fault, problem). It is not a *dōśaṃ* when it is done in the *puṟaṃ kaṭal* (outer sea). If they fish (vali) there, the fishes come to the coast (lit. 'step down to the east',

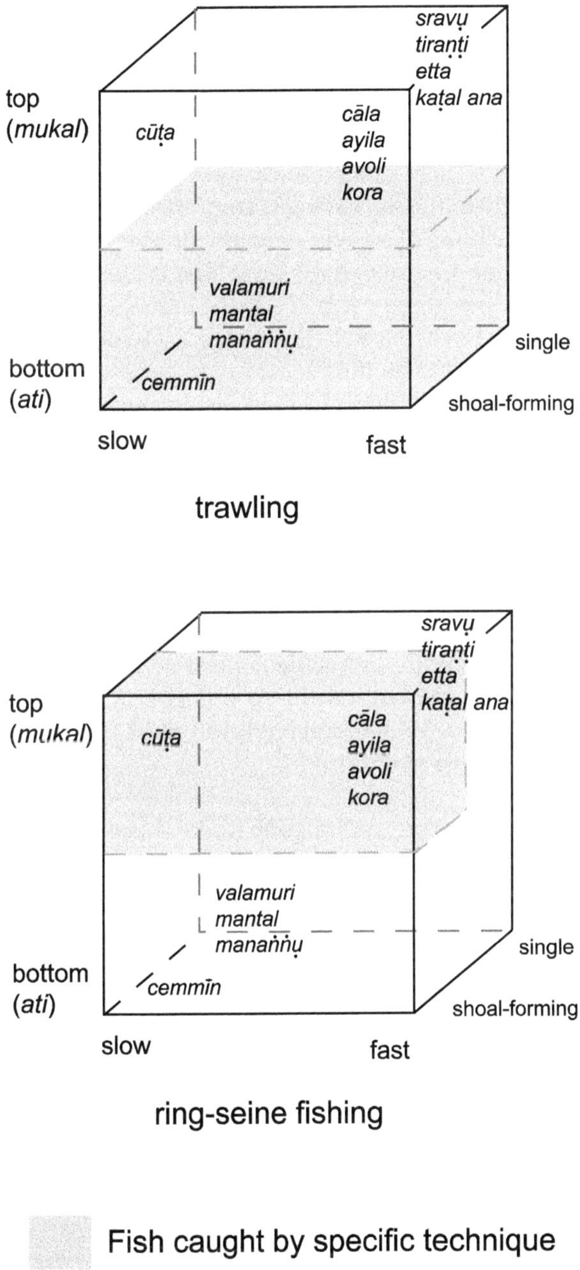

Figure 6.3: Bottom trawling and ring-seine fishing are perceived to affect different fish species.

mīn kilakkōṭṭu iraṅṅuka). They may not come to within 20 km [from the coast], if they come further than that to the east then we [the artisanal fishermen] will not get anything, that can be a *dōśaṃ*.

Just as easily the noise of trawlers may chase the fish away from the coast to the *puṟaṃ kaṭal,* where they cannot be caught by artisanal fishermen. In so doing they contribute to the sea having become 'restless' (*svastatayilla,* lit. 'not peaceful') due to the noise of an ever-growing number of boats, trawlers and ships. As Ramanathan Cherupuzhayil says: 'The convenience (*saukaryaṃ*) is not provided for the fish to come here [i.e. to the near-shore waters] any more'.

In Figure 6.3, I have marked the range in which bottom trawling and ring-seining (as done by the *vaḷḷaṃ*-s) are believed to have an effect. Indeed, just as the depths affected by these fishing techniques are mutually exclusive, the species that may be caught within them seem to be mutually exclusive as well. However, prawns are the exception, for they usually stay at the sea bottom and come to the surface occasionally. The ban on monsoonal trawling, even though of doubtful efficiency, is especially relevant to the catch of prawns, which are in the near-shore waters only during the south-west monsoon. As such it does seem to help artisanal fishermen, and is judged approvingly by them.

But that is not all. Without visible control of the catch trawling nets are dragged over the sea bottom. In so doing, fish eggs, juveniles and spawners are destroyed. Velayuthan Nediripil gives a vivid description of this indiscriminate effect of trawling:

Velayuthan Nediripil: You can imagine it like this: in my yard there is a small mango tree (*māvu*). It has got thousands of small, tender mangoes and many flowers. Two hundred ripe mangoes are also hanging in the tree. If we want to pluck these mangoes we have to do it with the greatest possible care. Tomorrow these flowers and tender mangoes will also become ripe mangoes. If we are asking the children to pluck the ripe mangoes now, they will do so, but thousands of flowers and tender mangoes will be lost. They will be lost because the children pluck the ripe ones in a barbarian manner (*apariśṭāmāyi poṭṭikkuka*). It is the same thing with the trawling. They may be getting a lot of fish, but many other fishes will perish (*naśippikkuna*). This way of fishing is wrong. Trawling is a wrong way of fishing.

Not only do the trawling nets catch fish indiscriminately, as Velayuthan explains, but what is more, the dragging of the trawling net over the sea bottom is known to destroy the fish eggs. Kunhiraman Kochikkattu comments:

Kunhiraman Kochikkattu: One kind of anchovy (*manaṅṅu̥*) is not seen [any more], it has about this size. (He shows the palm of his hand.) It was known as the 'red anchovy' (*coppan manaṅṅu̥*). This fish is totally absent nowadays ... There is another fish called *mullan* (silver belly), big *mullan*, in the size of my palm. It is not seen now. These are all fishes bursting from eggs. (He takes a leaf of coconut and continues.) Just like this the fish lay eggs, like a garland, and the whole will be laid over the mud beneath the sea. From this the young ones are coming up. When these trawling (*vali*) boats start moving through the areas of the eggs, the garland of eggs is broken up by the nets. Thus the eggs do not reach the maturity for hatching and hence they perish. This is my explanation in this case.

In talk about the decline of the catches, the spawning and maturation of young fish are rarely taken into account, perhaps because this subject is not part of the core conceptualisation of local environmental knowledge as presented in Chapter 4. The impact of the large nets with small mesh size (*telivu̥*) that are used by trawlers and purse-seiners is more easily imaginable.

Thus the effects of trawlers can be judged within several distinct frameworks: the depths affected, the impact of the engine noise, the dragging of the net over the sea bottom and its indiscriminate efficiency of catching. Fishermen involved in the *vaḷḷaṃ* units in Chamakkala know that, in recent years, it has become difficult to criticise trawlers explicitly, since some of the effects of trawling are shared by ring-seine fishing with *vaḷḷaṃ*-s. The generation of noise and the potentially harmful efficiency of large nets are two examples. As such, a critique of the 'others' may quickly turn into a critique of oneself.

Dams, Trawlers and the Mountain Water

From a Chamakkala viewpoint, implementing bans on the fisheries was the state's attempt to influence the west (i.e. the sea). Yet a kind of state intervention in the east, in the Western Ghats, has been perceived to influence the west even more profoundly. Remember that the arrival of the mountain water (*mala veḷḷaṃ*) at the beginning of the south-west monsoon has been considered as a central event of seasonal change in the sea. When the mountain water enters the sea, the latter is supposed to be rough for a few days, but subsequently its heat (in a figurative and literal sense) is supposed to be cooled (see Chapters 1 and 4). In addition, by entering the sea from the east, the mountain water generates a counter-current from west to east, which is supposed to carry along the fish of the *varśapeytụ* ('monsoon catch').

In the first two decades after India's independence the Kerala Water Authority began with the construction of large dams in Thrissur District for the generation of electricity and as reservoirs for irrigation and drinking water (Figure 6.4). Peechi dam was built between 1947 and 1957 (when it was in full storage), the Malampuzha and Unirakudi dams were constructed after 1957 and several smaller dams were built until the 1980s (CBIP 1987: 72–78, 1998: 19). These dams are now operated by the Kerala Department of Water Resources. Its engineers and administrators do not seem to have considered that these dams might have had a profound impact on the sea. Fishermen in coastal Thrissur District think differently, perceiving an disadvantage for three reasons. First, the near-shore sea remains 'hot' instead of being 'cooled' during the south-west monsoon. It has become unattractive for fish to come there. Secondly, the counter-current carrying along the fish of the *varśapeytụ* is absent. In consequence, fish feeding on fish of the *varśapeytụ* are missing in the east as well (Figure 6.5). Thirdly, the mud carried along with the mountain water may be nutritious and as such attractive for the fish, but is now lacking.

In order to situate discussions about the effect of the dams and the declining punctuality of the monsoon's arrival, let me turn to a conversation I recorded in early May 2000, a few weeks prior to the expected arrival of the south-west monsoon. Present were Kumaran Eranpurakkal, Gangadharan Kochikkattu and Gopalakrishnan Kochikkattu (three older fishermen in their sixties and seventies), Sathyan Kovil Tekku Valappil (thirty-seven, a guest worker on different boats), Suresh and myself. I had asked a question about the distribution of the *pankukaśụ* accounts (see Chapter 3), which was formerly done at the beginning of the monsoon, around 15 *iṭavaṃ* in the Malayalam calendar (about 1 June in the Western calendar).

Kumaran Eranpurakkal: *Pankukaśụ* means that we save one man's salary. Ten people move [to the sea], and from that one portion is saved, perhaps one tenth of the total amount [of the money earned at the auction]. At this time the monsoon also started, then we did not have the work.

Götz Hoeppe: Was the *pankukaśụ* done because you were not moving to the sea afterwards?

Kumaran, Gangadharan Kochikkattu, Sathyan Kovil Tekku Valappil: Yes.

Kumaran: Between 15 and 20 *iṭavaṃ* the sea certainly gets rough (*kaṭal iḷakkuṃ*).

Gangadharan: By 1 June the sea gets rough.

Gopalakrishnan Kochikkattu: Years ago the sea got rough on 17 *iṭavaṃ*, but after that it never happened at the correct time.

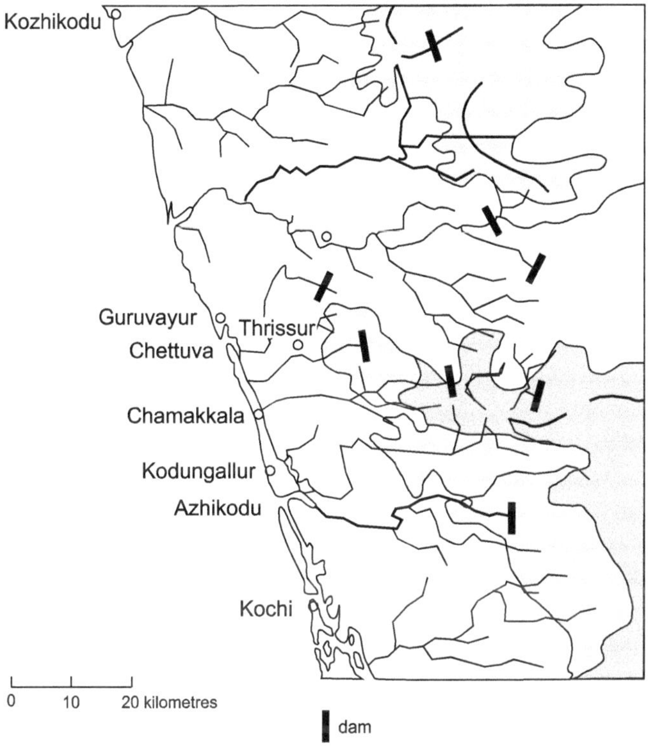

Figure 6.4: Map of central Kerala with dams built since Indian independence

Sathyan: That was a few years ago.

Gangadharan: Not a few years, that was maybe thirty years ago.

Sathyan: No, when I started to move to the sea the sea got rough on 15 *iṭavaṃ*. When I was working with Phalgunan's brother's boat they also had [i.e. kept] a *pankukaśu* [account]. That was for *vañci* people. I also got that money and I bought tapioca (*kappa*) and oil sardines (*cāla*). That was not thirty years ago, maybe it was twenty years ago. I still remember that on 15 *iṭavaṃ* the sea was getting rough and with the money from the *pankukaśu* I went to the Sri Murukan cinema and saw the Babu Mon film.[9]

Gangadharan: In the olden days the sea was getting rough on 15 *iṭavaṃ*, but now it has changed. Now the punctuality (*kṛtyaṃ*) has gone and the rain has also gone. From the *kuṃbhaṃ* month [mid-February to mid-March] itself the rain was starting. And in *makaraṃ* [mid-January to mid-February] there was rain as well.

Gopalakrishnan: The sea is not getting rough on 15 *iṭavaṃ*, and the reason for it is that that the dams have come and the trees have also been cut

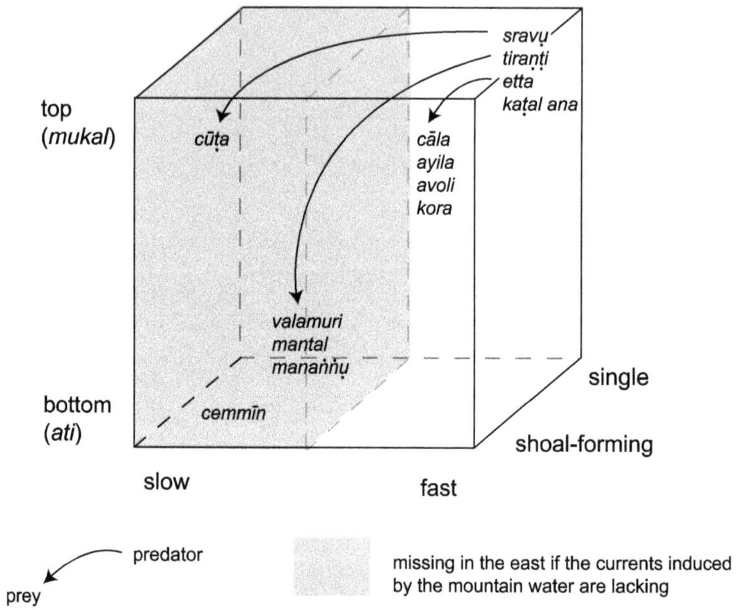

Figure 6.5: Scheme explaining the effects of the dams on the local fishery.

off. Previously, these trees stopped the clouds and thus they brought the rain. So many things have now turned around. If the mountain water (*mala veḷḷaṃ*) reaches the sea, then because of that excitement (*iḷakkaṃ*) the seawater will be agitated for three to five days. After that it will be peaceful. Then the fish come down [from the outer sea] and reach here, I mean, when the fresh water (*ñān veḷḷaṃ*) enters [the sea]. Nowadays pure water does not enter the sea. Instead of pure water, petrol and kerosene come into the sea today.

Gangadharan: So many dams (*ana*) have been constructed in the way of the rivers.

Gopalakrishnan: And the big trees are being cut and ministers are building houses with that wood. Usually sandalwood and teak wood are cut.

What started as a discussion about the *pankukaśụ* distribution quickly turned into a discussion about the lacking punctuality of the monsoon's arrival, the building of dams, deforestation, seawater pollution, immoral politicians and, implicitly, the present non-arrival of fish from the west. This is not an arbitrary enumeration of unrelated topics. Rather, they are related and go together as signs of the present 'time of progress' (*purōgamiccụ kālaṃ*). Implicitly, the lack of punctuality (*kṛtyaṃ*) of the roughening of the sea may be considered as a sign for the politicians' lack of truthfulness (*satyaṃ*). These terms are commonly related to each other.

Fishermen are not alone in perceiving an influence of the construction of dams on the sea. An engineer who was formerly employed by the Kerala Water Authority told me that because of the dams both the amount of the mountain water and its momentum have been reduced, thus expressing the fishermen's view in scientific terms. Yet most of the water stored in the dams is used for the generation of electricity and will eventually arrive in the sea.

Gopalakrishnan's comment that deforestation in the mountains has reduced the rainfall represents a common view, particularly of older people. The extent of deforestation in Kerala is not comparable to what has been observed in rural Rajasthan by Ann Gold and Bhoju Ram Gujar, who note that local Rajasthani villagers have blamed the (perceived) sparsity of rain on deforestation (Gold and Gujar 2002: 303f.). In Chamakkala, Gopalakrishnan mentions this effect to fill a gap in a causal chain that links the comfort of ministers (i.e. politicians) with the poverty of the fishermen: ministers have the trees cut → rain fails to be attracted (in addition: → dams hold up the remaining rain) → mountain water does not enter the sea any more → the sea is not getting rough as it did previously → the fish cannot come down from the outer sea (→ implicitly: fishermen do not catch fish). As such, it is a commentary on the neglect of the fisherfolk by politicians in power, mediated by the conceptualisation of environmental change. In this way, Chamakkala fishing is perceived to have been influenced significantly by (powerful) people who are not otherwise related to fishing (the *kiḻakkāran*-s, 'easterners', and the politicians).

While the construction of the dams in the east is perceived to have abolished the previously established regularity of the sea's roughening, the ban on monsoonal trawling and purse-seining has introduced a new regularity in the sea, i.e. in the west. When trawlers and purse-seiners commence operating in the outer sea (*puṟaṃ kaṭal*) after the end of the ban periods, they may chase the fish to the shore. This would happen in early August (for trawling) and early September (for purse-seining). If these boats chase fish towards the shore, this is regarded as a benefit locally.

Whether dams are constructed in the east or a government regulation affects the sea in the west, the fishermen perceive their environment to be altered by powerful forces that are inaccessible to them and remain exterior to their life-world. Understood as state intervention, these profound changes have been induced by people who do not engage in socially sanctioned relations with the sea. They affect and transform what has, in the fishermen's understanding, been hitherto perceived as innate.

Notes

1. The impact of the colonial state on the fisheries in Chamakkala dates back to the mid-nineteenth century. Exports of fish oil from Malabar District to Great Britain (shipped via the harbour in Kochi) began in the 1840s (Day 1865: xxi, xxv). These exports soon became sizeable, amounting to 4,140 tonnes over the five-year interval ending in 1863/64 (ibid.: xxi). However, during the entire nineteenth century there were no interventions that could match those initiated by the Madras Department of Fisheries after 1908.
2. Improved estimations of the exploitable fisheries resources of Madras Presidency were made in the early 1940s (Kurien 1985: A70).
3. See www.sophieprize.org (December 2005).
4. I thank the Deputy Director of Fisheries, Thrissur, for this information.
5. It is not surprising that access to moving *cākara*–s is more open than to fixed *cākara*-s like the one near Purakkad (Allapuzha District).
6. Kurien (1990: 478) mentions a demonstration of fishermen in front of 'the largest marine fishery research institute of India' (this is certainly the Central Marine Fisheries Research Institute, CMFRI, in Kochi; its scientists advise the state government on issues regarding the trawling ban). On this occasion, the following song was sung:

   You white-elephant scientists and researchers
   You servants of capitalism
   The research you conduct
   Is it to save the workers
   Or to serve the capitalists?
   Chorus: We are the children of the sea
   We know the secrets of the sea
   We don't need to be taught by anyone
   (Kurien 1990: 478; translated from Malayalam).

   One must suspect that this eloquent formulation was made up by experienced activists, rather than full-time fishermen.
7. I have taken the 1978-79 and 1988 numbers from PCO/SIFFS (1991b: 5) and the 1995 number from an article in *The Hindu* ('Panel findings on trawling questioned', Kerala edition, 31 July 2000).
8. During my twelve-month fieldwork in Chamakkala in 1999-2000 the only major political agitation by fishermen in Thrissur District was a large demonstration in Kodungallur against a decline in the subsidies for kerosene. Kerosene has become a major expense in the operation of the *vaḷḷaṃ* units.
9. Babu Mon was a well-known Malayalam actor in the 1980s.

7 Knowledge Change

The construction of dams in the Western Ghats and their perceived effect on the seasonal circulation and cooling of the sea as described at the end of Chapter 6 is a conceptual change that signifies an extension of the domain of human action and, conversely, an intervention of what is innate to the environment or 'natural'. With the world changed so profoundly and human–environment relations redefined, must not the fishermen's knowledge follow suit and be thoroughly transformed? The situation reminds one of Thomas Kuhn's (1962) concept of the historical development of science, in which periods of 'normal' science, characterised by problem-solving within the framework of an established paradigm[1], are interspersed by scientific revolutions in which a new paradigm is eventually adopted. Different paradigms are mutually incommensurable, there being no common language into which both can be translated without a loss of meaning. After a revolution, scientists live in a different world, writes Kuhn, and he describes the adherents of different paradigms as native speakers of different languages (1962: Chapter 10).

What, then, about the fishermen? What happens to their epistemic fabric (Appadurai 1989) in the contact with extraneous, ontologically incommensurable knowledge? Do they make syncretic accommodations of different knowledges (Cohen 1993; Piot 1999; Tsing 2005), or do they accept knowledge based on diverse ontologies coexisting side by side (Nichter 1980, 1992)? I close this book with some observations on the fishermen's usage of foreign, imported knowledges.

Government programmes and the import of new technology inevitably impose foreign knowledges on the fishermen. But, driven by practical interests as well as curiosity, fishermen are out themselves to discover new ideas about the world. Their sources are many. Most fishermen are literate and a number of them are avid readers of Malayalam newspapers such as *Mathrubhumi*, *Deshabhimani* and *Malayala Manorama*. Many people listen to the state radio regularly, not least for the weather forecast. Few fishermen watch TV regularly, but especially the younger ones often see movies in the two local cinemas. Experiences obtained by migrants to the

Arabian Gulf are important as well, and have affected the imagination of the outside world profoundly.

This unprecedented inflow of news, opinions, information and gossip is a key element of the present 'time of progress' (*purōgamiccu̥ kālam̥*, lit. 'progressed time'), which is contrasted to the 'old times' (pl. *paḻaya kālaṅṅaḷ*). Conspicuous consumption, new links to the outside world, electrification, water supply and toilets (financed by a French development project in the early 1990s), welfare plans of the Department of Fisheries and government reservations in education and employment, are as much part of the 'time of progress' as the introduction of new fisheries technology, the older fishermen's perception of a lack of respect of younger fishermen and the perceived decline of marine resources as well as the surmised lack of the truthfulness of the people, and perhaps even of the sea.

Imports are ambivalent, no matter if they are goods or information. This is particularly clear in regard to the import of cosmopolitan, Western medicine, which is announced as 'English medicine' in local pharmacies. Drawing on fieldwork data gathered in Karnataka, Nichter (1980) finds that cosmopolitan medicine is perceived as having an ambivalent power, particularly when contrasted with local medicine. The former is said to act rapidly and efficiently, though being of a vacillating and unstable nature. In contrast, local medicines are perceived as acting more smoothly but also more slowly and, perhaps, less efficiently (see Tsing 1993; Osella and Osella 2000; Gold and Gujar 2002). Similar statements could be made about foreign versus local technologies.

Becoming a Part of the World

During her fieldwork in Indonesia, in a provincial town in south Kalimantan, anthropologist Anna Tsing once sat together with a batik merchant late at night. This man gave her an elaborate account of his view that human action causes waves that reverberate in the atmosphere and influence other humans, and the power of the human heart to bring about changes in other humans and in the world more generally. Many anthropologists have encountered what Tsing (2005: 113) calls 'eclectic statements': apparently odd accounts of global warming, acid rain, nuclear radiation, faraway causes and reflections on their impact on local environments, local lives and local morals. Sometimes these statements are developed into systematic accounts that border on conspiracy theories, but in many cases they are explanatory nevertheless. They give reasons in a world of unknowing, a world of uncertainty. Not fitting into anthropological categories such as kinship, economy, religion or ritual, such statements

Figure 7.1: In the 'time of progress' (purōgamiccu̩ kālaṃ), boats may bear eclectic foreign names.

rarely find their way into published ethnographies. This is unfortunate, for such statements are often repercussions, immediate reactions to foreign concepts, which all too often remain alien (Figure 7.1). Sometimes, they reflect the attempt to discover the place of oneself in a world of global connections. Often, these statements take up the vocabulary of science, or what is perceived as such. But talking science means walking a tightrope. There is a chance to make oneself ridiculous with blunt statements about the connectedness of everything. Yet, at the same time, a skilled use of science, or of its vocabulary, may be strategically useful to signal an important social distinction: to be 'scientific' is not to be 'folk' (see Tarabout 1999). For the fisherfolk of Chamakkala, which has experienced social exclusion over many generations, this difference matters.

I shall confine myself to the 'eclectic' in accounts of environmental change and wonder about its meaning. Let us consider two examples. The first relates to the question of whether the *cākara*, the coastal mudbank, is related to the seasonal arrival of the mountain water at the beginning of the south-west monsoon. The second relates to a question about the perception of the lack of rain during the south-west monsoon in early August 2000. The first conversation took place between Shashi Kaniyaparambil, in his mid-forties a former Gulf migrant and now the owner of a *vaḷḷaṃ* unit, a *muṭuveṭṭi* and co-owner of a purse-seine boat, Suresh and myself:

Götz Hoeppe: Was the arrival of the mountain water good for the sea? Is there any relation to it [i.e. to the formation of the *cākara*]?

Shashi Kaniyaparambil: If you ask if it is good or not, then I would say that it is not so good. But some benefit is there and that is the pressure of the water (*veḷḷaṭṭinṟe valivụ*). This is the usual one, it is the water coming from the east to the sea [i.e. the mountain water]. In so many places the snow melts and joins the sea. The heat of the sea rises and the quantity of water increases as well. Scientific investigations (*śāstriyamāyiṭṭụ gavēśanaṃ*) have shown that the quantity of water in the sea has not decreased. It increases. From that we can understand that the snow mountains (*mannụ malakkaḷ*) melt and come here.

Suresh Muttichoor: Does the melted snow water come to Kerala?

Shashi: It may reach here through the outer sea (*puṟaṃ kaṭal*). Otherwise it is too hot here. Now there is even a scarcity of drinking water. A change has to happen to the sea, but it has not occurred yet.

Shashi does not give a straightforward answer to my question. After starting to answer, he turns to a general statement about the possible effect of 'melting snow mountains' upon the sea, a process that he believes affect Kerala eventually. In talking about the 'snow mountains' and the process of their melting, he is probably referring to global warming, that is, he is talking about the melting of Arctic or Antarctic icebergs or ice shelves and their expected influence on the rise of the sea level. Global topics like the greenhouse effect, environmental pollution, population growth and so on are featured in Kerala's Malayalam-language newspapers and in radio and television broadcasts. While sitting on the beach, Chamakkala fishermen, some of whom are avid readers of newspapers, occasionally discuss topics of this generality and debate their impact on the beach and the village.

This said, it will be less surprising that a question about the lack of rainfall during the south-west monsoon in early August 2000 should lead to concern about the long-term impacts of the 1991 Gulf war. What follows is part of a conversation between Vijayan Eranpurakkal and Kumaran Eranpurakkal (two fishermen in their fifties), Gangadharan Kochikkattu, Suresh and myself:

Götz Hoeppe: Why is the rain lacking?

Vijayan Eranpurakkal: If one wants to find out why there is no rain one has to do an expedition. Only somebody who has studied science (*śāstraṃ*) can know about it. If somebody without proper knowledge talks about it then it will be wrong. But some people do have a thought. So much kerosene (*manana*) is burnt in the sea. Then this kerosene spreads all over the sea. Perhaps this is the reason.

Kumaran Eranpurakkal: When the engines are blocked with dust then they are cleaned in the sea. While doing this, all petrol, kerosene and oil reaches the sea. It is done hundreds of times.

Gangadharan Kochikkattu: In the Gulf countries the ships are cleaned in the sea. This can take many days or even months.

Vijayan Eranpurakkal: And the oil is taken from the sea itself. What I want to say is that so much oil was spread in the sea. There was a war in Saudi Arabia or Muscat or Iran or Iraq. The oil tanks burst and the oil spread into the sea. That was four or maybe five years ago. The scientists said that this will make serious problems in India. I have heard that before.

It remains unclear to these men in what way oil spills in the Arabian Gulf might be linked to the local climate or to changes in the sea. Yet they do not doubt that, regardless of how far away these events might be, they are bound to have an impact in Chamakkala eventually. And it is in their having such an impact that these events connect the beach and the village to the world, or at least the world as it is portrayed in the media.

It is in their totality that such statements are meaningful. They suggest that the previously conceptualised, regular world (bounded, so to speak, by this conceptualisation) is enveloped in an uncertainty that is beyond comprehension. As such, the world has become too complex to be still understandable within the framework of previously established local conceptualisations: 'the old ideas are not working now' (Kunhiayyappan Kochikkattu). The envelope of uncertainty is the source of ambiguities and processes that one cannot influence and which one is forced to accept as part of the 'time of progress'.

Universal Standards

Uncertainty may prevail in the 'time of progress', but there is also the faith in universals, the faith in science, the faith in standards. The faith in standards of measurements, for example. Arjun Appadurai (1989: 178) notes that measurement is context-tied and that its changes, including the introduction of new units, may indicate shifts in the meaning of measuring itself. Valentine Daniel (1996: Chapter 3) draws attention to the violence imposed by new units and practices of measurement being introduced by the colonial government in Sri Lanka's tea production.

Shifts in units of measurement have occurred in the fisheries of Chamakkala. The universal, which means: Western, units of time, space and mass have entered the local world. All these coexist with local units: the Malayalam calendar is used parallel to the Western calendar, positions in the sea are expressed in terms of the *bhagaṃ* unit (water depth in arm spans) as well as in 'kilometres from the coastline' and the local *kottụ* (basket; corresponding to the English unit maund, i.e. about 37 kilograms) is still more common than the 'kilogram'.

There is a preference to use the Malayalam calendar for the timing of religious activities, while the Western calendar is used for most commercial purposes and bureaucracy, including government regulations. The date of the roughening of the sea at the beginning of the south-west monsoon, as well as the *pankukaśụ* distribution, are typically referred to in the Malayalam calendar (as the fifteenth day of the *iṭavaṃ* month). In contrast, the dates of the trawling ban(s) are always linked to the Western calendar. As such, a regularity of the environment, which was previously expressed by a local unit of timekeeping, is lost, but a government-regulated seasonality, which is linked to a global unit of timekeeping, has been introduced, further underlining the alienness of the impact on the sea of trawlers and their regulation by the government.

The shift from *bhagaṃ* to kilometre as a measure for the distance from the beach invites a speculation about the conceptualisation of marine space. Since previously the depth of the sea and the distance from the beach were denoted by the same unit of depth, one could surmise that the spatial structure of the sea was largely cognised as a scale of depth. The recent introduction of denoting distances in kilometres may have added another dimension and made a more precise description of this space possible.

New Metaphors

Given this book's emphasis on figurative language, the invention of new metaphors suggests itself as an expression of knowledge change. If metaphor, most generally, means to see the less familiar in terms of the more familiar, then it is not surprising that local society, commonly experienced in local lives, serves as a source of metaphors. This is demonstrated by metaphors and similes used to describe the *cākara*'s unsteady behaviour: 'Today the *cākara* moves with the *mantravādi*-s ['sorcerers']' and 'Nowadays the *cākara* is [behaves] like a politician (*cākara ippōḻatte rāśṭriyakāreppōle*)'. Both the works of sorcerers and the doings of politicians are considered as unpredictable, unreliable and able to be influenced with money. Thus expressed, the *cākara*, previously understood as a particularly pure and truthful part of the environment, has entered the low grounds of morality in the 'time of progress'. Sociality with the environment is re-evaluated.

Cleansing Myths

I opened this chapter with a note on the shifting meaning of the 'innate' as it is precipitated by the construction of the dams in the Western Ghats. Yet, if I compare the spectrum of explanations and utterances concerning the impact of the mountain water on the sea, from the myth retold by Krishnankutti at the beginning of Chapter 1 to the conversation on the irregularity of the monsoon's onset at the end of Chapter 6, I realise that the basic concept is retained (mountain water reaching the sea being necessary for assuring catches near the coast), while the moral connotations of the myth have been abolished. The myth has been cleansed from what was not deemed reasonable any more, much in the way which the historian Paul Veyne documented in his study of Greek belief in their myths (1988). This cleansing is a way of liberating local knowledge (often circumspectly) of 'fallacies', which are perceived because new knowledge, often ontologically distinct, enters the local world. Broadly speaking, it is a metaphorical process in the sense of Max Black: an old conceptualisation is looked at through the screen of current beliefs with its remainders being removed or questioned.

Cleansing is noticeable with regard to other concepts as well, the Lotus Sea being one example. The existence and the properties of this distant part of the sea, allegedly covered by lotus leaves during the hot season and therefore a cherished destination for the fishes' recreation, are a matter of debate (see Chapter 4). Chandran Kozhissery, in his mid-sixties, a former fisherman and now middleman (*tarakan*), explains:

Götz Hoeppe: What is the Lotus Sea?
Chandran Kozhissery: The Lotus Sea is known as the Seventh Sea, but there is no clear evidence (*rēkha*). So many people say that the Lotus Sea is there. But one cannot support [the claim] that it is there officially (*itụ audhyōdika māyiṭṭụ uṇṭō eṇụ samar̠tikkal par̠r̠illa*).

Doubt about the existence of the Lotus Sea is present even in the older generation. While Krishnankutti is convinced of its existence, claiming that some people have actually seen it, his age-mate Gangadharan disagrees: 'We have seen the end of the sea (*ā kaṭalinr̠e aṭṭaṃ oke kaṇdetti*). Now we know that there is no Lotus Sea'. Having lived in South India all his life, Gangadharan cannot have seen the 'end of the sea' himself. But many younger men did: as migrants to the Arabian Gulf they have taken an aeroplane to reach the other side of the Arabian Sea. Some of them have watched the sea attentively through the aeroplane window. But none had seen a sea covered with lotus petals. Ramanathan Cherupuzhayil, who worked in Saudi Arabia before returning to fishing, does not feel that this observation challenges the existence of the Lotus Sea. He says:

Ramanathan Cherupuzhayil: Sometimes the *ayila* (mackerel) of the *tulaṃ* month has got this smell, hasn't it? In *ciṅṅaṃ, kanni* and *tulaṃ* [mid-August to mid-November] we catch this fish in the west. Then we move to the place where *ayila* lives. Previously we did not go beyond 10 *bhagaṃ*, that means 10 kilometres ... When we did not have any other fish then we would move there to put the net. The fishes from that area have a special smell (*pratyēkata vāsana*) of lotus (*tāmaru*), so it was called the Lotus Sea. Other than that there is no other sea there.

With the background of his own experiences, Ramanathan 'rationalises' the concept of the Lotus Sea. He does not disprove the idea of a Lotus Sea but reduces, or indeed changes, its conceptualisation to a version that is consistent with empirical observation.

From Literality to Figurativeness

The import of information often raises doubt about established conceptualisations, but it may also confirm them in unsuspected ways. Let us consider the rise of the sea towards the west, above the beach level (Chapter 4). Always understood as a surprising fact contradicting the everyday experience of water flowing down, this claim has been commonly explained with the relation of truthfulness between the sea and the coastal dwellers.

Keshavan, a retired fisherman in his sixties, remembers as a 'proof' for the claim that the sea rises towards the west a newspaper article that described the discovery of the wreckage of an Indian aeroplane in the Atlantic Ocean by American remote-control submarines. He remembers precisely that the wreckage was found at a depth of 6,634 feet. Ignorant of intermediate continents, he concluded that the increase in the height of the sea must continue at least to this value. A search in the Internet confirmed Keshavan's memory. In June 1985 an Indian Airways aeroplane crashed 110 miles west of Ireland in what was the worst terrorist attack worldwide prior to 11 September, 2001. The wreckage was found at a depth of 6,700 feet.[2]

Venu Irinjathuruththy (forty) and Vijayan Kovil Tekku Valappil (forty-three), both active fishermen on *vaḷḷaṃ*-s, disagree with the claim supported by Keshavan:

Götz Hoeppe: When the fish have moved to the west you say: 'gone up up' (*kayaṟi kayaṟi pōyi*).
Vijayan Kovil Tekku Valappil: Yes. We will say 'gone up' (*kayaṟi pōyi*).
Götz: Some people say that the sea is standing (*nilkkunattu*) above the land.
Vijayan: That is not true. We have to consider what the people from

here say and what science says (*nātiluḷḷa ālkāṟanmar parayunatuṃ śāstriyamāyittuḷḷatuṃ*). If we climb up some stairs then we can surely say that the height of this land is increasing. That is scientific (*atụ śāstriyamāttānụ*) ...

Venu Irinjathuruththy: We are not able to talk about that. It is difficult to say that the sea is higher than the earth. We cannot say so.

Vijayan: Water flows down. If this [land] here is below the sea, then the water must flow here.

Götz: Some people say that the sea does not flow here because of its truthfulness (*satyaṃ*).

Vijayan: (Laughs.) Ha, ha. (More seriously:) That is true in one way. But what if we think in another way?

Vijayan does not outrightly deny the rising, but argues against it by referring to the authority of a scientific viewpoint. Like all fishermen, Vijayan and Venu use the verb *kayaṟuka* in describing the movement of fish towards the west. However, in regarding the sea as flat, their use of *kayaṟuka* is metaphorical. The older fishermen had used this verb literally.

The most critical opinion about the rising of the sea can be expected from fishermen who have been migrant workers in the Arabian Gulf. Ramanathan Cherupuzhayil (mid-fifties), now a shareholder on a *vaḷḷaṃ*, is one of the few men who went back to fishing after working as a labourer in Saudi Arabia. He says:

Ramanathan Cherupuzhayil: In the west (*paṭiññāṟụ*) [the sea] is high and in the east (*kiḻakkụ*) it is low. It is standing (*nilkkukayānụ*) like this. The Arabian Sea (*aṟabi kaṭal*) has no right (*avakāśaṃ*) to pursue anything over here [i.e. flow over here]. But then, in the Gulf it is different. There it goes down towards the west, but here it goes up towards the west.

It seems that Ramanathan's view of the structure of the sea has indeed been affected by his stay in the Arabian Gulf, not in the way of doubting the rise of the sea, but by adding that in the Gulf the sea goes down in the west. He makes the old view consistent with his new experiences. In saying that the sea has no (traditional) right (*avakāśaṃ*) to move towards the east, Ramanathan hints at a contract or agreement that seems to be established between the sea and the people. Thus, the moral notion of truthfulness (*satyaṃ*) appears to be maintained in local discourse, but it is not 'allowed' to interfere with rationalistic concepts of impersonal forces, according to which water flows down. The split of the (perceived) ontological domains of physical reality and morality is realised here by moving from a literal to a figurative understanding of marine space.

We are witnessing here a movement of the border of figurative and literal speech. It is the inertia of language as implied by the continued usage of some terms that makes it possible to pursue a 'conceptual archaeology' (Gentner et al. 2001: 210), a project for which Mary Hesse, the philosopher of science, gives useful advice:

> An expression initially metaphoric may become literal (a 'dead' metaphor), and what is at one time literal may become metaphoric (for example the Homeric 'he breathed forth his life', originally literal, is now a metaphor for death). What is important is not to try to draw a line between the metaphoric and the literal, but rather to trace out the various mechanisms of meaning-shift and their interactions. (1980: 116–117)

Translation

Translating the vocabulary of science, or what is perceived as such, into local conceptualisations is common, and may be considered a special case of the 'cleansing' described above. One day, Phalgunan Thaykattu, a man in his late forties, who has spent eighteen years working on fishing ships in the Arabian Gulf, explained to me the formation of plankton and its role as the food of fishes:

Phalgunan Thaykattu: At the time of rain this salt water and fresh water will *mix*. From it the substance called *plankton* is formed. It happens in the middle depth of the sea (*kaṭalinṟe iṭattaṭṭu*). But then it moves to the wide bottom [of the sea] (*aṭipparappu̥*). When the heat of the sun recedes, when it has been raining for three or four months, this substance (*sādanaṃ*) will settle down ... I mean the substance called *plankton*. This *plankton* is produced in the rainy season (*varśa kālaṃ*). The *plankton* is produced when the waters are joining (*kuṭiyōjikkumbol*). It stays in the top of the sea (*kaṭalinṟe mukal*). *Cemmīn* comes up in order to eat it. In the months *dhanu̥* and *makaraṃ* [mid-December to mid-February] the plankton moves down, then *cemmīn* should be caught from the bottom. Now it comes from the *pelagic*. Therefore, it [*cemmīn*] eats the food (*bhakśanaṃ*) from that place. We [humans] also move to the places where food (*bhakśanaṃ*) is available …

[To know] this is called the cleverness of the *fisheries science* (fisheries science–*inṟe miṭukku̥*). The *fisheries science* is okay. The ones who wrote the *fisheries science*, between that and what we see there is no difference (*vitiāsaṃ*).

Phalgunan's account closely resembles the usual explanation, described in Chapter 4, of how a mixture of waters brings about the formation of *tari* ('small fragment, seed'), the food of fish – except for his introduction of three English terms: *mix*, *plankton* and *pelagic*. Acknowledging the insight of the 'fisheries science', he uses the noun *plankton* as a synonym of *tari*, the adjective *pelagic* side by side with its synonym *kaṭalinṟe mukal*, and the verb to mix next to its Malayalam equivalent *kuṭiyōjikkumbol*. By doing so he has poured the old wine of 'local knowledge' into the new bottle of (global) 'science'.

Being Scientific

In the cases considered so far, *śāstraṃ* is understood in a rather encompassing way, more defined by contexts of foreignness and power than by its epistemology or semantic content. For example, Shashi and Vijayan, when talking about 'melting snow mountains' or 'kerosene spreading all over the sea' in the Arabian Gulf name scientific investigations (*śāstriyamāyiṭṭu gavēśanaṃ*) and *expeditions* as sources of knowledge. In doing so, they do not refer to the *śāstraṃ* tradition of Hindu scriptures (which the Malayalam term also designates; see below), but to a corpus of knowledge and a way of knowing which they perceive as foreign (indicated by the reference to an English term) and powerful (see Tsing 1993: 30, 257; Gold and Gujar 2002: 290).

The emergence of science's power on the local scene has brought along a reassessment of the status of local cultural categories. Consider astrology (called *jyoti śāstraṃ*, '*śāstraṃ* of light', in Malayalam), which, according to its practitioners, is scientific beyond any doubt.[3] Indeed, the standard kind of astrological diagnosis, the *prasnaṃ*, is perceived as an investigation. In its performance, cowrie shells are drawn, their number is interpreted, handbooks and almanacs full of rows of numbers are consulted, calculations are made and diagrams are drawn. Though incomprehensible for untrained onlookers, these procedures may resemble the conduct of a scientific investigation. Several astrologers have purchased computers, and these are always prominently displayed to their clients. These computers link the astrologer's work in a direct way with modern and powerful technology. Yet this does not mean that these computers are actually used: an astrologer from a neighbouring village has glued a postcard with a *yantraṃ* (ritual diagram) onto the screen of his (broken) monitor. Clients of *mantravādi*-s commonly justify their resort to *mantravādaṃ* ('sorcery') by pointing out that it has been proved to be scientific (*śāstriyamāyi*). A category that is itself perceived as extraneous is here used to justify the resort to local, indigenous activities. And thus, while resorting to *mantravādaṃ* may be a reaction to powerful foreign influences, it is itself subjugated to the judgement of what is perceived as foreign knowledge.

Even though 'science' and its related terms 'fisheries science', *gavēśanaṃ* (research, investigation) and *gavēśanaṃkāran* (investigator, scientist) are strongly associated with the 'time of progress', this does not mean that the link of knowledge with foreignness and power is new. In the past, *śāstraṃ* itself, defined as 'scriptural knowledge', was linked to Brahmanical dominance by virtue of its definition. The Vedas, scriptures that are closely linked to Brahmanical ritual practice and Brahmins' claim to high social status, are considered as *śāstraṃ* par excellence. As seen in Chapter 5, Brahmanical *mantravādaṃ*, involving *mantraṃ*-s recited in Sanskrit, is thought to be particularly powerful. From this point of view, 'science' continues a pre-existing tradition of linking power with foreignness.

Parallel Knowledges

The border between the figurative and the literal often remains blurred, keeps shifting and depends on individual speakers. In general, middle-aged men are more open-minded towards new, i.e. global, technology than the older ones. Their speech may not differ much from that of the elders, but they understand statements as metaphorical which to some older men express literal truths. Sharing a common language facilitates communication, regardless of whether individuals consider a specific utterance as literal or figurative.

When such instances occur more often, their totality may indicate that extraneous, global knowledge, itself being understood literally, has pushed local conceptualisations into the domain of the figurative. The notion of the rising of the sea provides us with one example. Other cases are less evident, such as the notion of *kaṭalamma*, or, more generally, the motherliness of the sea. Fishermen, who do not hesitate to make technical statements about new engines, mobile phones, GPS satellite position devices or their own migration experience to the Arabian Gulf talk, in other contexts, talk quite naturally about the sea's truthfulness and its rising toward the west as expressions of a personal, moral relatedness with the fishermen. This context sensitivity would seem like another hint at the persistence of the local. But how long can a 'figurative local' and a 'literal global' coexist?

The growing criticism of notions such as the rising of the sea towards the west among middle-aged and younger fishermen reminds one of Max Weber's notion of the 'disenchantment of the world' in which superempirical beings are no longer believed to affect the empirical world causally (Weber 1920). The selective use of figurative language makes it possible to speak (and reflect) about the two domains of 'truth' at the same time, even in the same sentence, given the consensus of speakers and hearers about the border between literal and figurative language. Therefore, even speech itself may be 'enchanted' and 'disenchanted' at the same time, depending on the understanding of the border between the literal and the figurative.

The strategical use of essentialised, figurative language in political discourse is another example of a parallel 'enchantment' and 'disenchantment' of the sea. At a time when, in Chamakkala, the sea's truthfulness is challenged and the notion of *kaṭalamma* is regarded by several fishermen as merely metaphorical (hinting at a 'disenchantment'), political fisheries activists in Kochi and Thiruvananthapuram have used the notion of *kaṭalamma* in their agitation against monsoonal trawling and the operation of foreign fishing ships. It is largely due to the popular movie *Cemmīn* (1965) that the notion of *kaṭalamma* is known in Kerala far beyond the fishing community (see Chapter 4). As noticed in Chapter 6, the notion of the children of the sea is extensively used in the agitation of Christian activists from south Kerala. This is remarkable, since the very same activists introduced scientific research into the debate on the trawling ban. While Chamakkala fishermen do not subscribe to this reversal of *kaṭalamma* as the 'sea mother', it is semantically well defined, emotionally meaningful and inclusive for Hindu, Muslim and Christian fishermen alike. This insistence on an exclusive kin-like relationship with the environment could be termed a re-enchantment of the sea. Political discourses about fisheries involve people from a number of non-fishing communities. They have different views of the sea and will not equivocally share an understanding of the 'sea as mother' or the fisherfolk as the 'children of the sea' in a literal sense. Only as metaphors do these statements evoke the fisherfolk's connectedness with the sea, which people from the inland lack. The notion of the artisanal fisherfolk being the children of the sea may be a creation for the activists' strategic purposes, but it is also a reminder of the fact that the sea cannot be understood by scientific reasoning alone.

Notes

1. According to Okasha (2002: 81), a Kuhnian paradigm consists of two components: '[F]irstly, a set of fundamental theoretical assumptions that all members of a scientific community accept at a given time; secondly, a set of "exemplars" or particular scientific problems that have been solved by means of those theoretical assumptions, and that appear in the textbooks of the discipline in question'.
2. See www.flight182.com and en.wikipedia.org/wiki/Air_India_Flight_182 (January 2006).
3. Demarcating astrology and science has proved to be a difficult task for philosophers; see Curd and Clover (1998: Chapter 1). Turner (1995: 108) calls astrology 'scientific non-science'.

Appendix A
Glossary

aḻi	river mouth
andhaviśvāsaṃ	superstition
antarīkṣaṃ	weather, atmosphere
aṟappa	small stream
arāyi	small water inlet
arivụ	knowledge
aryakkāran	skipper, helmsman
ati	bottom, usually sea bottom
ayittaṃ	pollution, untouchability
bhagaṃ	depth measured in *marụ*
bhakśanaṃ	food
bhanḍāraṃ	money collection box
BJP	Bharatiya Janata Party (Hindu nationalist party)
boat	trawling boat
brāhmarakṣassụ	shrine for the spirits of deceased Brahmins
cākara	coastal mudbank
cāḷi	mud
caiṭanyaṃ	a deity's life force, consciousness
cāppa	fish curing yard
carrier	boat for transporting a catch
cāṭṭukkāran	'jumping man' (in ring-seine fishing operation)
ceḷi	mud
cella	mud
cēṟụ	mud
chit fund	savings club
CMFRI	Central Marine Fisheries Research Institute (Kochi)
CPI	Communist Party of India
CPI(M)	Communist Party of India (Marxist)
CPI(ML)	Communist Party of India (Marxist-Leninist)

daivaṃ	god
dēvan	deity
dīvara sabha	lit. 'assembly of the clever ones', caste organisation of fishermen
dōṣaṃ	problem, fault (both astrological and in normal life)
double net	mini trawl, a small trawling net
EEZ	Exclusive Economic Zone, zone of exclusive rights to use marine resources, extends 200 nautical miles (371 kilometres) into the sea
ēṭaṃ	high tide (rising of water)
eitihyaṃ	legend
fisheries science	perceived corpus of foreign, Western knowledge of fish and fishing
gavēśanaṃ	research
iḷakkaṃ	movement, shaking
jāti	'caste'
jīvan	life, soul
jōli	work, business, employment
kāla vaṟṣaṃ	south-west monsoon
kālavastaṃ	weather
kamarụ	fluorescent glow (of seawater and fish)
kanpa vala	beach-seine net
kara	shore
kara kaṟṟụ	wind from land to the sea
karmaṃ	magic spell
kāṟṟụ	wind
kaṟutta vāvụ	new moon
kaṭakkoṭi	headman of the Araya caste
kaṭal	sea
kaṭalamma	sea mother
kaṭappuṟaṃ	seashore
kaṭavụ	beach (section of the shore)
kāvụ	sacred grove
kayal	backwater
kiḻakkāran	'person of the east'
kiḻakkụ	east

kiḻakkuppuṟaṃ	'outside in the east'
kolli vala	boat-seine net
kollụ	current, flow of water
kōlụ	wind
kōpaṃ	anger, rage of the sea
kṣōbhaṃ	agitation, turbulence, fright
kula toḻil	caste occupation (lit. 'lineage work')
kūṭṭaṃ	fish shoal
lakṣyaṃ	'sign' for fish
lōkaṃ	world
lōpụ vala	gill net
lunki	colourful loincloth
madrasa	Quranic school
maḻa	rain
mala veḷḷaṃ	mountain water
Malayali	a speaker of Malayalam, resident of Kerala
mantraṃ	magical spell
mantravādaṃ	'sorcery', black magic, conjuration
mantravādi	ritual expert, 'sorcerer'
marụ	length unit (one arm span, i.e. about 1.7 metre)
Matsyafed	Kerala state fisheries fisheries cooperative apex organisation
matsyatoḻilāḷi	fisherman, lit. 'fish worker'
mīn	fish
muṇṭụ	monochromatic loincloth
muṟukkān	betel, areca nut, etc. used for chewing
muṭuveṭṭi	small plank-built boat
nakṣatraṃ	star, constellation
nērca	vow offering
nīrụ	current
nīti	justice, righteousness
nīṭụ vala	gill net
nīṭụ vali	gill net fishing
niyamaṃ	rule
oḻukkụ	flow, current
paḻaya kālaṃ	old time
pani	(manual) work (as opposed to *jōli*)
paṇi	fever

pankukaśu̇	joint savings account of *vañci* fishermen
paṭiññāṟu̇	west
paṭiññāṟuppuṟaṃ	outside in the west
PCO	Programme for Community Organisation (Thiruvananthapuram)
pelappu̇	shoal of fish
poḷa	toxic, coloured water
poṭi	small grain, powder, dust
prakṛti	'nature', constitution
prasnaṃ	astrological diagnosis
pūja	Hindu ritual
pūjari	Hindu priest
pūḷa	pollution
punnūlu̇	sacred thread
puṟaṃ kaṭal	outer sea
purōgamiccu̇ kālaṃ	'time of progress'
puruśan	man
ring vala	(encircling) ring-seine net
RSS	Rasthriya Swayamsevak Sangh (Hindu nationalist youth organisation)
śabdaṃ	noise
śakti	power, strength
śāstraṃ	scriptures, law, rule
satyaṃ	truth, truthfulness, reality
saukaryaṃ	convenience
srānku̇	engine coordinator on a *vaḷḷaṃ*
strī	woman
tali	pivot for *taṇṭu̇* (oar)
tāmaru̇ kaṭal	Lotus Sea
taṇṭu̇	oar
taraku̇	commission
taṟavāṭu̇	land–house–family unit, coll. ancestral home
tari	grain, food of fish
taṭakaṃ	lake, lagoon calm area in the sea
teḷivu̇	mesh size
tīradēśaṃ	coastal area
tōṇi	boat
tula vaṟṣaṃ	north-east monsoon

ulpādanaṃ	production, procreation
utsavaṃ	festival, celebration
vala	net
valakkāran	lit. 'net man', worker on a *vaḷḷaṃ*
vali	coll. trawling
vali boats	trawling boats
vaḻi paṭu	offering (to a temple)
vali vala	trawling net
vaḷḷaṃ	plank-built boat
vañci	dugout boats in varying size (for two, three or five men)
varśapeytu̥	catch of the south-west monsoon
vaṭi vala	stick net
vāvu̥	phase of moon
veḷḷaṃ	water
veḷutta vāvu̥	full moon
vēnal	summer
vēnal pakal	hot season
veyil	sunshine, sun heat
vikr̥ti	vicious attitude, mischief
vīśu̥ vala	cast net
yantraṃ	ritual diagram
yukti	tactics, cleverness

Malayalam Months

ciṅṅaṃ	mid-August to mid-September
kanni	mid-September to mid-October
tulaṃ	mid-October to mid-November
vriścikaṃ	mid-November to mid-December
dhanu̥	mid-December to mid-January
makaraṃ	mid-January to mid-February
kumbhaṃ	mid-February to mid-March
mīnaṃ	mid-March to mid-April
mēṭaṃ	mid-April to mid-May
iṭavaṃ	mid-May to mid-June
mithunaṃ	mid-June to mid-July
karkkaṭakaṃ	mid-July to mid-August

Appendix B
A Household Census

Figure B.1: Family portrait: Babu Kovil Tekku Valappil, his wife Leela, daughter Sajitha, son Sajith and some of their friends

I use statistical material from the 1991 census and a survey of 326 households that I conducted in 2000 in order to situate the fisherfolk of Chamakkala in their social environment and study the local impact of Gulf migration.

At the 1991 census, there were 5,085 households with a population of 27,457 in the *panchayat* (measuring approximately seventeen square kilometres). At that time, Chamakkala, the *panchayat*'s seaside ward, had a population of 2,308 in 394 households.[1] The *panchayat* mirrors the diversity of Kerala's population.[2] East of the National Highway there are several households of *savarna*[3] ('caste') Hindus: Brahmins, Ambalavasis (temple servants), Vishwakarmas (artisans) and Nayars (warriors and landowners). The majority of the Hindu population are *avarna* ('casteless') Hindus: Izhavas ('toddy tappers'), Kaniyan (astrologers), Mannan (washers), Pulayar, Thandan, Vettuva (agricultural labourers) and Valan (backwater fishermen). The small fraction of Roman Catholic Christians lives east of the National Highway as well. The Muslim fraction increases west of the National Highway and is largest in Chamakkala near the coast. Except for a few Nayar households, only *avarna* Hindu castes and Muslims live near the sea.

My household census covers the area west of the mosque, reaching somewhat into the neighbouring wards of the two adjacent *panchayat*-s. Most of the households of the fishermen who work at the Chamakkala beach (*kaṭavu*) are situated here (Figure B.1). The questionnaire was aimed at gathering information about the residents (number, family relations, education, occupation, relation to fishing), their caste (*jāti*) or community (*sammudāyaṃ*[4]), their housing (ranging in standard from thatched huts to air-conditioned villas), landownership and standard of living (as expressed in the possession of material goods like bicycles, motorcycles, TV sets, refrigerators, fans, etc.). Tables B.1 and B.2 give an overview of some census results. The 326 households in my survey have a total population of 1,903 persons (including migrants who are temporarily absent). The residents belong to four communities: Muslims (54 per cent) and Hindu Araya (33 per cent), Izhava (8 per cent) and Vettuva (5 per cent).

With approximately 2,290 persons per square kilometre the population density is high, being three times the all-Kerala average. This is typical for coastal villages (see Kurien 2000a: 184). The census reveals that fishing-related work and work as migrants in India or abroad account for one-third each of male primary occupations, the remainder being a variety of occupations from agricultural labour to high government employment (including a retired *tahsildar*[5]). An occupational differentiation of the male members of the four communities is noticeable (Table B.1). Fishing and fish selling are practically restricted to Arayas and Muslims: none of the Izhava men is engaged therein and there is only one Vettuva fisherman and one Vettuva *cappa* worker.[6] The majority of fishermen belong to the Araya caste (105 of 112), while the majority of fish vendors are Muslims (33 of 38). Likewise, the other occupations relating to fishing, i.e. work in a *cappa* (fish curing yard), as an auctioneer or a *tarakan* (middleman, moneylender), are dominated by Muslims and Arayas. The majority of the working Araya male population in Chamakkala is engaged in fishing

Table B.1 Male primary occupations in Chamakkala.

Occupation	Muslim	Araya	Izhava	Vettuva	Total
Fishing	6 (2.2%)	105 (53.6%)	0	1 (2.9%)	112 (20.4%)
Fish vending	33 (12.2%)	5 (2.6%)	0	0	38 (6.9%)
cappa work (fish curing yard)	6 (2.2%)	0	0	1 (2.9%)	7 (1.3%)
Fish auctioneer (employed)	0	2 (1.0%)	0	0	2 (0.4%)
Middleman (*tarakan*)	0	5 (2.6%)	0	0	5 (0.9%)
Agricultural/manual labour	12 (4.4%)	3 (1.5%)	9 (18.8%)	22 (62.9%)	46 (8.4%)
Skilled factory work	3 (1.1%)	2 (1.0%)	0	0	5 (0.9%)
Technical employment	11 (4.1%)	6 (3.1%)	3 (6.3%)	2 (5.7%)	25 (4.6%)
Petty trade (incl. shop employment, cook)	35 (13.0%)	10 (5.1%)	13 (20.8%)	2 (5.7%)	60 (10.9%)
Coir fibre production	1 (0.4%)	0	9 (18.8%)	0	10 (1.8%)
Farming	4 (1.5%)	1 (0.5%)	1 (2.1%)	0	6 (1.1%)
Lower government employment	1 (0.4%)	4 (2.0%)	1 (2.1%)	1 (2.9%)	7 (1.3%)
Teacher	0	1 (0.5%)	1 (2.1%)	0	2 (0.4%)
Quran teacher in *madrasa*	1 (0.4%)	0	0	0	1 (0.2%)
Astrologer/*pūjari*	0	2 (1.0%)	1 (2.1%)	0	3 (0.5%)
Military and police	0	1 (0.5%)	2 (4.2%)	0	3 (0.5%)
High government employment	0	3 (1.5%)	0	0	3 (0.5%)
Shop owner	8 (3.0%)	2 (1.0%)	2 (4.2%)	0	12 (2.2%)
Big trade	5 (1.9%)	0	0	0	5 (0.9%)
Migrated in India (unskilled)	5 (1.9%)	3 (1.5%)	0	2 (5.7%)	10 (1.8%)
Migrated in India (skilled/professional)	1 (0.4%)	1 (0.5%)	0	0	2 (0.4%)
Migrated to the Gulf (unskilled)	120 (44.4%)	27 (13.8%)	4 (8.3%)	4 (11.4%)	155 (28.2%)
Migrated to the Gulf (skilled/professional)	18 (6.7%)	13 (6.6%)	2 (4.2%)	0	33 (6.0%)
Total	270	196	48	35	549

Source: census conducted between 30 May and 9 August 2000.

Table B.2 Housing conditions and Gulf migrants in Chamakkala.

Community	Number of households	Persons	Persons per household	Houses: percentage of huts	Houses: percentage of concrete buildings	Houses: percentage electrified	Gulf migrants per household
Muslim	155	1,035	6.7	3	46	94	1.02
Araya	123	627	5.1	15	17	80	0.36
Izhava	32	152	4.8	3	0	81	0.19
Vettuva	16	89	5.6	19	0	56	0.31

Source: census conducted between 30 May and 9 August 2000

or fish-related work (59.7 per cent). Only 16.7 per cent of working Muslim men are occupied in this field. On the other hand, more than half of the working Muslim male population (53.3 per cent) is working as migrants in other parts of India and in the Persian Gulf. Among working Araya men, the fraction of migrants is much smaller (22.4 per cent) and among Izhava and Vettuva men from this area it is even less (12.5 per cent and 17.1 per cent, respectively). Most Izhava men near the beach work as agricultural labourers, in the production of coir fibres or in petty trade[7], while most Vettuva men work as agricultural labourers or do head-load work.[8]

Notes

1. I thank the Director of Census Operations, Thiruvananthapuram, for providing me with these unpublished data.
2. There are no official statistics available that list the *panchayat* population according to religion or caste.
3. The distinction of *savarna* and *avarna* Hindu castes first appears in the Rig Veda (Fuller 1992: 12ff). Brahmans (priests), Kshatriyas (warriors), Vaishyas (people) and Shudras (servants) are *savarna* ('caste') Hindus. Fishing castes, castes of agricultural labourers etc. do not belong to any of these categories and are therefore termed *avarna* ('casteless') Hindus. A peculiarity of the caste system in Kerala is the lack of Kshatriyas and Vaishyas among *savarna* castes (Fuller 1976: 33-38).
4. While *jāti* (lit. 'species') formally denotes Hindu castes as differentiated according to their occupation, *sammudāyaṃ* applies to 'communities' more generally (and more ambiguously). The latter includes Muslims and Christians. While conducting the census, I realised that members of low-status castes (e.g. Vettuva) felt denigrated when being asked about their *jāti*, but did not mind me asking about their *sammudāyaṃ*. Their answer would be the same.
5. A *tahsildar* is the revenue subcollector and often the head official of a *taluk* administration.
6. A *cappa* is a fish curing yard. Fish are salted and stored for later consumption. Most *cappa*-s have been shut down due to the widespread use of ice for cooling fish. Fish from *cappa*-s is sold at a much lower rate than fresh fish or fish cooled with ice.
7. For more information about the Izhava caste, see Ananthakrishna Iyer (1909: 277–341), Thurston (1909: II, 392–418) and Osella and Osella (2000).
8. For more, though in many ways outdated, information about the Vettuva caste, see Ananthakrishna Iyer (1909: 128–134) and Thurston (1909: VII, 394–404).

Appendix C

West and East in the 'Time of Progress'

Figure C.1: An import from the east: cricket on the beach.

Throughout this book I have made reference to the 'time of progress' (*purōgamiccu kālaṃ*, lit. 'progressed time'), the present period, which Chamakkala's fishermen associate with new contacts with the outside world and new cultural imports as well as declining morals and dwindling catches (see Figure C.1). Construing with it the 'old times' (pl. *paḻaya kālaṅṅaḷ*), all time is divided into these two categories. It a totalising concept. In the 'time of progress' the meanings that had previously been associated with the cardinal points east and west, the 'inside' of the village and the 'outside' world of the east (see Chapter 2), are challenged and transformed.

In recent years, the gap between the fishermen and the 'people of the east' has become more permeable. Today, Arayas may become 'people of the east' (*kiḻakkāran*-s). By taking up work inland some local men and women have done so. Self-declared 'advanced' Arayas keep emphasising that they not only work in the inland but rarely, if ever, visit the beach, for lack of social contacts (with those they regard as lower-status fishermen). By assimilating the habits and styles of dress (men: white *muṇṭụ*-s and white shirts, women: fine *sari*-s) of the east, fishermen regard them as 'people of the east' (albeit without granting them higher status). The fraction of local Araya men working outside the fisheries is now at about forty per cent (see Appendix B, Table B.1). It was certainly considerably lower when fishing (and fishing-related work) were considered as the almost exclusive 'lineage work' of the Arayas in the past.

New Aspirations

In many ways fishing, the 'work of the west', is still considered a thrilling experience. The present, 'progressive' fishing (*purōgamana maṭsya baṇdhānaṃ*) makes use of outboard motors and motorised winches. The use of this equipment and the responsibility for its maintenance, are a source of fascination and self-assurance for younger fishermen. It is acknowledged (by young and old) as having removed much of the hardship that fishing was previously associated with. Much of the old men's excitement, enthusiasm and 'appetite' for fishing is shared by young men. This is noticeable especially when the men embark on a fishing trip after several idle days. Older men comment upon the younger men's fascination with (and reliance on) machines while also commenting on their lack of bodily strength.

When I asked young Araya men about their plans for the future, most expressed their wish to be employed outside fishing. Their principal argument against fishing is the irregular income. Only three months per year are said to provide good landings. In contrast, jobs in an office or some other business are perceived to provide a regular income. Yet the attractiveness of the *kiḻakkupuṟaṃ*, where these jobs are located, extends far beyond the idea of establishing economic stability. It is the attraction to a world where foreign goods are available for consumption and where movies from Bombay or abroad may be seen. It is a world beyond the narrow confines and the social control experienced by living near the beach.

Most of the young fishermen have tried to find employment in the east, the *kiḻakkuppuṟaṃ*, after leaving school. But not all the work in the *kiḻakkuppuṟaṃ* is *jōli* (esteemed white-collar work). Young Araya men, who have often failed in high school before passing SSLC, the tenth-grade examination, are more likely to find work as painters, as salesmen or as money collectors in a chit fund (collective savings group). Their hopes and expectations of working in the *kiḻakkuppuṟaṃ* rarely turn into professional success. Experiences of professional failure are common for young (and mostly unmarried) Araya men, but most of them continue to be attracted to Chentrappini Centre, which all fishermen regard as being part of the *kiḻakkuppuṟaṃ*. On a borrowed bicycle, often shared by three young men, they go to the cinema or they just hang out near the highway, watching the bustling life while drinking a *soda* (locally bottled carbon dioxide-enriched water). The movies seen in the two local cinemas have influenced their ideas of modernity. They are a source of up-to-date music and fashion.

Owning Land in the East

Despite the failure of many young fishermen to find employment in the east, it is clear that, in principle, it is possible to 'appropriate' the east and to become 'eastern'. Participating in Gulf migration is regarded as the most promising way to do so. This is exemplified by returning migrants, who, in some cases, have literally managed to become *kiḻakkāranmar* ('easterners') by buying land (and, in some cases, residing) there. On the one hand, due to the high population density of the coastal area, it is inevitable that one has to go further inland in order to find land for sale. On the other hand, however, land in the east is often regarded as being of a supreme quality. This quality is largely inferred from the caste status of previous residents. For example, land previously owned by high-status caste Nayars has a superior quality to land that was previously owned by low-status Izhavas. The reasoning behind this valuation is the widely shared concept that land and people are related by a substantial circulation of qualities: the living live from the land in which their ancestors are buried. Living on land that was previously inhabited by high-status people is understood to entail absorbing some of these people's substantial quality by ingesting the food grown on this land, as well as being physically present there (see also Uchiyamada 1995, Osella and Osella 2000: 146-151).

An example from Chamakkala is provided by Muhammad (pseudonym). Coming from a previously impoverished Muslim family, he succeeded in accumulating considerable wealth as one of the village's first migrants to the Persian Gulf. Continuing to reside on West Tippu

Sultan highway (where he has built an American-style two-storey villa with marble floors and air conditioning), he recently bought three acres of land from a meanwhile impoverished Nayar family. He did not buy the sacred grove (*kāvu̧*) on this land for fear of Nagas (snake deities) and Hindu spirits residing therein (see Uchiyamada 1995). Besides owning this plot, he also owns several paddy fields in the near-coastal inland. Other families from Chamakkala, usually those with Gulf migrants, have bought paddy fields as well, including those of all the Araya *tarakan*-s whom I questioned about this. Reasons to own these fields were the assurance of an autonomous food supply, the quality (taste) of the rice and the wish to own land, still regarded as somewhat unusual for fishermen who previously owned small plots at the seashore only.

Visiting the East

If owning land inland may seem like a maximal appropriation of the 'east' (*kil̲akkuppur̲aṃ*), visiting it for a day may be regarded as its opposite: minimal appropriation. Regardless of their age, when going to the *kil̲akkuppur̲aṃ*, Chamakkala fishermen dress well: they wear freshly washed and ironed shirts and *muṇṭu̧*-s (monochromatic loincloths). While older men rarely go there, married, middle-aged fishermen visit the *kil̲akkuppur̲aṃ* in order to get paperwork done in an office, buy specific items (perhaps for the boats) or go to a bank. For a few, the bar in Edamuttam, Chentrappini's neighbouring village to the north, is the destination for enjoyment. It sells beer and foreign ('global') liquors, in contrast to the toddy (*kallu̧*, palm wine) shop near the beach which sells only locally produced toddy.

The white cotton *muṇṭu̧*-s and shirts worn in the offices and in the bar contrast with the old shirts or cheap, colourful synthetic *lunghi*-s or shirts that are worn at work or in the toddy shop at the beach. Says one man: 'People wear new shirts in the *kil̲akkuppur̲aṃ*. After two months these shirts are worn in the *paṭiññār̲u̧* and new shirts are stitched [for the *kil̲akkuppur̲aṃ*]'. This observation is indicative of the change in attitude towards fishing that is noticeable among younger and some middle-aged fishermen: fishing as an occupation is their second, if not their last, choice.

The East Affects the 'Work of the West'

Most parents do not blame their sons for thinking in this way. In fact, there are few of them who want their sons to become fishermen. However, as far as fishing itself is concerned, older and middle-aged fishermen complain about the negative influence of the east on the conduct of young fishermen. Thus, the *aryakkāran* of a Chamakkala *vaḷḷaṃ* explains that they spend much of the hard-earned savings there quickly. He claims that, for young fishermen, earning money that can be spent in the *kiḻakkuppuṟaṃ* has become their main reason to work. But then, perhaps due to last night's show in the cinema, they arrive late at the beach in the morning and thus delay the fishing trip. They are said to lack discipline towards the older ones and do not show the right attitude towards the sea. Singing popular film songs in the boats, they break the respectful silence which the older fishermen deem necessary. Young men who have worked in the east before joining a fishing team may be blamed for lacking practice and may be accused of making mistakes.

The perception of older and middle-aged men that the young ones lack the respectful attention to the work of fishing is widespread. A middle-aged fisherman comments: 'Wearing chappals (flip-flops) they [the young ones] are stepping into the seawater nowadays! They must take their chappals off! They should not step onto the net with their shoes. They [the old ones] are not crossing the net with their legs'. My observations on the beach in Chamakkala confirm that younger men prefer to wear flip-flops in on the beach – and step on the nets wearing them. Statements about the younger fishermen are usually more differentiated than that. While the younger ones' lack of respect and devotional attitude is commonly commented upon, a former Gulf migrant who returned to Chamakkala fisheries, criticises the younger men's lack of interest in earning money from fishing:

Dasan Eranpurakkal: The fishermen need to have the temper to go to the sea at any time. But our men do not have this enthusiasm (*jyaraṃ*) to work any more. They do not feel like making money from the sea by hard work. When a small catch has been made they go home, take a wash and go to the *kiḻakkuppuṟaṃ* well-dressed until their money is finished. In some other beaches (sing. *kaṭavụ*) the fellows are working day and night and thus they make a lot of money. They never move to the *kiḻakkuppuṟaṃ*. What money they get will remain with them. But our people move to the *kiḻakkuppuṟaṃ* and enjoy the city life, like drinking alcohol and going to the movies. That is what they do until their money is finished. In their mind there is no place for studying the sea and making money from it.

Manoharan Eranpurakkal echoes this viewpoint and notes that, whether in fishing or in working in an office, responsibility and sincerity (*ātmārttata*) are necessary. Older fishermen may consider the comparison of fishing and office work as inappropriate, but for younger men, who value white-collar work (*jōli*) more than manual work (*pani*), the reverse may be implied. Dasan and Manoharan describe fishing in terms of the common perceptions of the 'work of the east': modern and organised, with the potentiality of gaining economic benefits. This point of view is echoed by several former Gulf migrants who have, like Dasan, invested large amounts of money into boats and gear. I am not sure if Dasan and Manoharan would claim that upward social mobility can be achieved while remaining in fishing. In the end, it is this question which many younger fishermen ask themselves.

These contested views of the 'outside in the east' and the 'west' illustrate the increasing permeability towards occupations other than fishing. Young Arayas, for whom upward social mobility seems possible, feel attracted towards the east and tend to see it in positive terms: in many ways it is synonymous with the 'time of progress'. In this way they seem to follow a trend that is defined by the 'advanced' Arayas and many Muslims. Older Araya men, on the other hand, feel the permeability as a threat, arguing that negatively valued attitudes and behaviour are entering and spreading in the west. These may even pollute the sea and disturb the relation of truthfulness. This point of view may not be surprising. After all, these older men cannot expect to be able to participate in upward social mobility. In defending the old values of the *paṭiññāṟu* they are defending a realm of their own that is closely linked to their identity. Thus, older men tend to revel in memories of the olden times (pl. *paḻaya kālaṅṅaḷ*): when truthfulness prevailed, men were strong and sharks were abundant in the sea. As such, the more the 'west' and the 'outside in the east' permeate each other, the more the older ones shift from space (east/west) to time (old time/time of progress) as constituents of their belonging, constituents of what is 'their own'. Space and time, if looked at in this way, are ideological constructs. The 'west' and the 'outside in the east', the 'old times' and the 'time of progress' set out an ambiguous field of contrasting associations, in which personal itineraries are questioned and challenged like never before.

Bibliography

Ahmad Kutty, E.K. 1995. 'The Mujahid Movement and Its Role in the Islamic Revival in Kerala'. In: Asghar Ali Engineer (ed.), *Kerala Muslims: A Historical Perspective*. New Delhi: Ajanta Publications, pp. 68–82.

Ananthakrishna Iyer, L.K. 1909. *The Cochin Tribes and Castes*. Vol. I. Madras: Higginbotham.

Ananthakrishna Iyer, L.K. 1912. *The Cochin Tribes and Castes*. Vol. II. Madras: Higginbotham.

Anon. 1915. 'Papers from 1899 Relating Chiefly to the Development of the Madras Fisheries Bureau', *Madras Fisheries Bulletin*, 1.

Appadurai, Arjun. 1989. 'Transformations in the Culture of Agriculture'. In: C.M. Borden (ed.), *Contemporary Indian Tradition: Voices on Culture, Nature, and the Challenge of Change*. Washington: Smithsonian Institution Press, pp. 173–189.

Appadurai, Arjun. 1996. *Modernity at Large: Cultural Dimensions of Globalization*. Minneapolis: University of Minnesota Press.

Astuti, Rita. 1995. *People of the Sea: Identity and Descent among the Vezo of Madagascar*. Cambridge: Cambridge University Press.

Astuti, Rita, Gregg E.A. Solomon and Susan Carey. 2004. *Constraints on Conceptual Development: A Case Study of the Acquisition of Folkbiological and Folksociological Knowledge in Madagascar.* Monographs of the Society for Research in Child Development, vol. 69 (3). Boston and Oxford: Blackwell.

Atran, Scott. 1990. *Cognitive Foundations of Natural History: Towards an Anthropology of Science*. Cambridge: Cambridge University Press.

Atran, Scott. 1998. 'Folkbiology and the Anthropology of Science: Cognitive Universals and Cultural Particulars'. *Behavioural and Brain Sciences* 21: 547–609.

Babadzan, Alain. 1993. *Les Dépouilles des dieux: Essai sur la religion tahitienne à l'époque de la découverte*. Paris: Maison des Sciences de l'Homme.

Babb, Lawrence A. 1975. *The Divine Hierarchy: Popular Hinduism in Central India*. New York: Columbia University Press.

Bal, D.V. and K.V. Rao. 1990. *The Marine Fisheries of India*, revised edition. New Delhi: Tata McGraw-Hill.

Balakrishnan Nair, N. 1991. *Report of the Fisheries Expert Committee on Ban on Monsoon Trawling in Kerala*. Trivandrum: Government of Kerala.

Balan, V. 1984. 'The Indian Oil Sardine Fishery: A Review'. *Marine Fisheries Extension Series, Technical and Extension Series*, Kochi 60: 1–10.

Barth, Fredrik. 2002. 'An Anthropology of Knowledge'. *Current Anthropology* 43: 1–18.

Basso, Keith H. 1990. *Western Apache Language and Culture: Essays in Linguistic Anthropology*. Tucson: University of Arizona Press.

Basso, Keith H. 1996. *Wisdom Sits in Places: Landscape and Language among the Western Apache*. Albuquerque: University of New Mexico Press.

Bavinck, Maarten. 2001. *Marine Resource Management: Conflict and Regulation in the Fisheries of the Coromandel Coast*. New Delhi: Sage Publications.

Beck, Brenda E.F. 1969. 'Colour and Heat in South Indian Ritual'. *Man* (NS) 4: 553–572.

Beck, Brenda E.F. 1976. 'The Symbolic Merger of Body, Space and Cosmos in Hindu Tamil Nadu'. *Contributions to Indian Sociology* 10: 213–243.

Best, Elsdon. 1909. 'Maori Forest Lore'. *Transactions of the New Zealand Institute* 52: 433–481.

Bhimachar, B.S. and P.C. George. 1950. 'Abrupt Set-Backs in the Fisheries of the Malabar and Kanara Coasts and "Red Water" Phenomenon as their Probable Cause'. *Proceedings of the Indian Academy of Sciences, Section B* 31: 339–350.

Bird-David, Nurit. 1990. 'The Giving Environment: Another Perspective on the Economic System of Hunter-Gatherers'. *Current Anthropology* 31: 183–196.

Bird-David, Nurit. 1992. 'Beyond the "Original Affluent Society": A Culturalist Reformulation'. *Current Anthropology* 33: 25–47.

Bird-David, Nurit. 1999. '"Animism" Revisited: Personhood, Environment, and Relational Epistemology'. *Current Anthropology* 40: S67–S91.

Black, Max. 1962. *Models and Metaphors: Studies in Language and Philosophy*. Ithaca: Cornell University Press.

Black, Max. 1979. 'More about Metaphor'. In: Andrew Ortony (ed.). *Metaphor and Thought*. Cambridge: Cambridge University Press, pp. 19–43.

Bloch, Maurice. 1991. 'Language, Anthropology and Cognitive Science'. *Man* (NS) 26: 183–198.

Bloch, Maurice. 1992. 'What Goes Without Saying: The Conceptualization of Zafimaniry Society'. In: Adam Kuper (ed.). *Conceptualizing Society*. London: Routledge, pp. 127–146.

Bloch, Maurice. 1998a. *How We Think They Think: Anthropological Approaches to Cognition, Memory and Literacy*. Boulder: Westview Press.

Bloch, Maurice. 1998b. 'Why Trees, Too, Are Good to Think With: Towards an Anthropology of the Meaning of Life'. In: Laura Rival (ed.). *The Social*

Life of Trees: Anthropological Perspectives on Tree Symbolism. Oxford / New York: Berg, pp. 39–55.

Boyarin, Jonathan. 1994. 'Space, Time, and the Politics of Memory'. In: Jonathan Boyarin (ed.). *Remapping Memory: The Politics of TimeSpace*. Minneapolis: University of Minnesota Press, pp. 1–37.

Boyer, Pascal. 1996. 'What Makes Anthropomorphism Natural: Intuitive Ontology and Cultural Representations'. *Journal of the Royal Anthropological Institute* (NS), 2: 83–97.

Braune, W. 1960. 'Abd al-Kaadir'. In: *Encyclopedia of Islam* (New Edition), Leiden: Brill, vol. I, pp. 69–70.

Busby, Cecilia J. 2003. 'We are the Sea's Children: Environmental Politics and Discourses of Rights in the Fishworker's Movement in Kerala'. Talk at the conference *People and the Sea II*, MARE Centre for Maritime Research, Amsterdam, 6 September 2003.

Caldwell, Sarah. 1999. *Oh Terrifying Mother: Sexuality, Violence and Worship of the Goddess Kali*. New Delhi: Oxford University Press.

Calvino, Italo. 1997. *Invisible Cities*. Translated by William Weaver. London: Vintage.

Carey, Susan. 1985. *Conceptual Change in Childhood*. Cambridge, Mass: MIT Press.

CBIP 1987. *Large Dams in India*, Vol. I. Publication No. 197. New Delhi: Central Board of Irrigation and Power.

CBIP 1998. *Large Dams in India*, Vol. II. Publication No. 197. New Delhi: Central Board of Irrigation and Power.

Cohen, Anthony P. 1993. 'Segmentary Knowledge: A Whalsay Sketch'. In: Mark Hobart (ed.). *An Anthropological Critique of Development: The Growth of Ignorance*. London: Routledge, pp. 31–42.

Curd, Martin and J.A. Clover. 1998. *Philosophy of Science: The Central Issues*. New York / London: W.W. Norton.

Dale, Stephen F. 1980. *Islamic Society at the South Asian Frontier: The Mappila Muslims of Malabar, 1498–1922*. Oxford: Clarendon Press.

Dale, Stephen F. and M. Gangadhara Menon. 1995. 'Nercca: Saint-Martyr Worship among the Muslims in Kerala'. In: Asghar Ali Engineer (ed.), *Kerala Muslims: A Historical Perspective*. New Delhi: Ajanta Publications, pp. 174–199.

Daniel, E. Valentine. 1984. *Fluid Signs: Being a Person the Tamil Way*. Berkeley: University of California Press.

Daniel, E. Valentine. 1996. *Charred Lullabies: Chapters in an Anthropography of Violence*. Princeton: Princeton University Press.

Davidson, Donald. 1979. 'What Metaphors Mean'. In: Sheldon Sacks (ed.). *On Metaphor*. Chicago: University of Chicago Press, pp. 29–45.

Day, Francis. 1863. *The Land of the Perumauls, or Cochin, its Past and Present*. Madras: Gantz Brothers.

Day, Francis. 1865. *The Fishes of Malabar*. London: Bernard Quaritch.
D'Cruz, Titto. 1998. *The Ring Seine: Evolution and Design Specifications*. Thiruvananthapuram: South Indian Federation of Fishermen's Societies.
Descola, Philippe. 1994. *In the Society of Nature: A Native Ecology in Amazonia*. Cambridge: Cambridge University Press.
Descola, Philippe and Gísli Pálsson. 1996. *Nature and Society: Anthropological Perspectives*. London and New York: Routledge.
Devanesan, D.W. 1943. 'A Brief Investigation into the Causes of the Fluctuation of the Annual Fishery of the Oil-Sardine of Malabar, Sardinella Longiceps, Cuv. & Val., Determination of its Age and Account of the Discovery of its Eggs and Spawning Ground'. *Madras Fisheries Bulletin*, 28: 1–38.
Dossi, Beatrice. 1998. *Samen, Seele, Blut: Die Zeugungstheorien des alten Indien*. Munich: Akademischer Verlag.
Eck, Diana. 1998. 'The Imagined Landscape: Patterns in the Construction of Hindu Sacred Geography'. *Contributions to Indian Sociology* (NS) 32: 165–188.
Ellen, Roy. 1993. *The Cultural Relations of Classification. An Analysis of Nuaulu Animal Categories from Central Seram*. Cambridge: Cambridge University Press.
Ellen, Roy and Holly Harris. 2000. 'Introduction'. In: Roy Ellen, Peter Parkes and Alan Bicker (eds), *Indigenous Environmental Knowledge and its Transformations: Critical Anthropological Perspectives*. Amsterdam: Harwood Academic Publishers, pp. 1–33.
Engineer, Asghar Ali. (ed.). 1995. *Kerala Muslims: A Historical Perspective*. New Delhi: Ajanta Publications.
Feldhaus, Anne. 1995. *Water and Womanhood: Religious Meanings of Rivers in Maharashtra*. New York: Oxford University Press.
Fernandez, James W. 1988. *Persuasions and Performances: The Play of Tropes in Culture*. Bloomington: Indiana University Press.
Flood, Gavin. 1996. *An Introduction to Hinduism*. Cambridge: Cambridge University Press.
Frohnmeyer, L. J. 1979 [1913]. *A Progressive Grammar of the Malayalam Language*. 2nd edition. New Delhi: Asian Educational Services.
Fuller, C.J. 1976. *The Nayars Today*. Cambridge: Cambridge University Press.
Fuller, C.J. 1992. *The Camphor Flame: Popular Hinduism and Society in India*. Princeton: Princeton University Press.
Gabriel, Theodore. 1996. *Hindu-Muslim Relations in North Malabar, 1498–1947*. Lewiston: Edwin Mellen Press.
Gell, Alfred. 1988. 'Technology and Magic'. *Anthropology Today* 4: 6–9.
Gell, Alfred. 1998. *Art and Agency: An Anthropological Theory*. Oxford: Clarendon Press.

Gentner, D., B. Bowdle, P. Wolff and C. Boronat. 2001. 'Metaphor is like Analogy'. In: D. Gentner, K.J. Holyoak and B.N. Kokinov (eds), *The Analogical Mind: Perspectives from Cognitive Science*. Cambridge, Mass.: MIT Press, pp. 199–253.

George, M.J. 1988. *Study of Shrimp Trawling in the South West Coast of India, Particularly Kerala*. Thiruvananthapuram: Programme for Community Organisation.

Gigerenzer, Gerd. 2002. 'The Adaptive Toolbox'. In: G. Gigerenzer and R. Selten (eds). *Bounded Rationality: The Adaptive Toolbox*. Cambridge, Mass.: MIT Press, pp. 37–50.

Godelier, Maurice. 1999. *The Enigma of the Gift*. Translated by Nora Scott. Cambridge: Polity Press.

Gold, Ann Grodzins and Bhoju Ram Gujar. 2002. *In the Time of Trees and Sorrows: Nature, Power, and Memory in Rajasthan*. Durham: Duke University Press.

Gopinathan, C.K. and S.Z. Qasim. 1974. 'Mud Banks of Kerala – Their Formation and Characteristics'. *Indian Journal of Marine Sciences* 3: 105–114.

Government of Kerala. 1983. *The Kerala Marine Fishing Regulation Acts and Rules 1980*. Thiruvananthapuram: Department of Fisheries, Government of Kerala.

Govindan, V. 1916. 'Fishery Statistics and Information, West and East Coasts, Madras Presidency'. *Madras Fisheries Bulletin*, 9.

Gudeman, Stephen and Alberto Rivera. 1990. *Conversations in Colombia: The Domestic Economy in Life and Text*. Cambridge: Cambridge University Press.

Gundert, Hermann. 1995 [1872]. *A Malayalam and English Dictionary*. Kottayam: D.C. Books.

Gupta, Akhil. 1995. 'Blurred Boundaries: The Discourse of Corruption, the Culture of Politics and the Imagined State'. *American Ethnologist* 22: 375–402.

Hallowell, A. Irving. 1960. 'Ojibwa Ontology, Behaviour and World View'. In: S. Diamond (ed.), *Culture in History: Essays in Memory of Paul Radin*. New York: Columbia University Press, pp. 19–52.

Hesse, Mary. 1980. *Revolutions and Reconstructions in the Philosophy of Science*. Brighton: Harvester Press.

Hesse, Mary. 1993. 'Models, Metaphors and Truth'. In: F.R. Ankersmit and J.J.A. Mooij (eds), *Knowledge and Language*, Vol. III: Metaphor and Knowledge. Dordrecht: Reidel, pp. 49–66.

Hill, Jane H. and Bruce Mannheim. 1992. 'Language and World View'. *Annual Review of Anthropology* 21: 381–406.

Hirschfeld, Lawrence A. and Susan A. Gelman. 1994. *Mapping the Mind: Domain Specificity in Cognition and Culture*. Cambridge: Cambridge University Press.

Hornell, James. 1910. 'Report of the Results of a Fishery Cruise along the Malabar Coast and to the Laccadive Islands in 1908'. *Madras Fisheries Bulletin* 4: 71–126.

Hornell, James. 1920. 'The Origins and Ethnological Significance of Indian Boat Designs'. *Memoirs of the Asiatic Society of Bengal* 7: 139–256.

Hornell, James. 1931. 'A New Protozoan Cause of Widespread Mortality among Marine Fishes'. *Madras Fisheries Bulletin* 11: 53–66.

Hornell, James. 1937. 'The Fishing Methods of the Madras Presidency, II. The Malabar Coast'. *Madras Fisheries Bulletin* 27: 1–68.

Hornell, James and M. Ramaswami Nayadu. 1924. 'A Contribution to the Life-History of the Indian Sardine, with Notes on the Plankton of the Malabar Coast'. *Madras Fisheries Bulletin* 17: 129–197.

Howell, Signe. 1996. 'Nature in Culture or Culture in Nature? Chewong Ideas of 'Humans' and Other Species'. In: Philippe Descola and Gísli Pálsson (eds), *Nature and Society: Anthropological Perspectives.* London: Routledge, pp. 127–144.

Hutchins, Edwin. 1995. *Cognition in the Wild.* Cambridge, Mass: MIT Press.

ICAR. 1971. *Report on Sardine and Mackerel Resources*. Delhi: Indian Council of Agricultural Research.

Induchoodan, V.T. 1969. *The Secret Chamber: A Historical, Anthropological and Philosophical Study of the Kodungallur Temple*. Thrissur: Cochin Devaswom Board.

Ingold, Tim. 2000. *The Perception of the Environment: Essays on Livelihood, Dwelling and Skill*. London: Routledge.

Isaac, T.M. Thomas. 1997. 'Economic Consequences of Gulf Migration'. In: K.C. Zachariah and S. Irudaya Rajan (eds), *Kerala's Demographic Transition: Determinants and Consequences*. New Delhi: Sage Publications, pp. 269–309.

Jackson, Michael. 1989. *Paths toward a Clearing: Radical Empiricism and Ethnographic Inquiry*. Bloomington: Indiana University Press.

Jeffrey, Robin. 1992. *Politics, Women and Well-Being: How Kerala became 'a Model'*. London: Macmillan.

Keil, Frank C. 1989. *Concepts, Kinds and Cognitive Development*. Cambridge, Mass.: MIT Press.

Kirfel, Willibald. 1968 [1920]. *Die Kosmologie der Inder nach den Quellen dargestellt*. Hildesheim: Georg Olms.

Klausen, Arne Martin. 1968. *Kerala Fishermen and the Indo-Norwegian Pilot Project*. Oslo: Universitetsforlaget.

Krech, David and Crutchfield, Richard S. 1948. *Theory and Problems of Social Psychology.* New York: McGraw-Hill.

Kuhn, Thomas. 1962. *The Structure of Scientific Revolutions*. Chicago: University of Chicago Press.

Kuhn, Thomas. 1977. *The Essential Tension: Selected Studies in Scientific Tradition and Change*. Chicago: University of Chicago Press.

Kumar, Deepak. 1995. *Science and the Raj, 1857–1905*. New Delhi: Oxford University Press.

Kumaran, M., T.M. Yohannan and Feroz Khan. 1988. 'Marine Fish Calendar: 3. Calicut'. *Marine Fisheries Extension Series, Technical and Extension Series* 81: 1–9.

Kurien, John. 1980. *Fisheries Co-operatives in Kerala: A Critique*. Madras: FAO/BOBP.

Kurien, John. 1985. 'Technical Assistance Projects and Socio-Economic Change: Norwegian Intervention in Kerala's Fisheries Development'. *Economic and Political Weekly* 20: A70–A88.

Kurien, John. 1990. 'Knowledge Systems and Fishery Resource Decline: A Historical Perspective'. In: W. Lenz and M. Deacon (eds), *Ocean Sciences: Their History and Relation to Man. Deutsche Hydrographische Zeitschrift, Ergänzungsheft*, Reihe B, 22: pp. 476–480.

Kurien, John. 2000a. 'The Kerala Model: Its Central Tendency and the "Outlier"'. In: G. Parayil (ed.), *Kerala: The Development Experience: Reflections on Sustainability and Replicability*, London: Zed Publications, pp. 178–197.

Kurien, John. 2000b. *Factoring Social and Cultural Dimensions into Food and Livelihood Security Issues of Marine Fisheries: A Case Study of Kerala State, India*. Working Paper No. 299. Thiruvananthapuram: Centre for Development Studies.

Kurien, John. 2000c. 'Capacity, Costs and Output Changes in the Small-scale Fisheries Sector of Kerala State'. Unpublished mimeo.

Kurien, John and T.R. Thankappan Achari. 1988. 'Overfishing along Kerala Coast: Causes and Consequences.' *Economic and Political Weekly* 25: 2011–2018.

Kurien, John and A.J. Vijayan. 1995. 'Income Spreading Mechanisms in Common Property Resource: Karanila System in Kerala's Fishery'. *Economic and Political Weekly* 30: 1780–1785.

Kurien, John and Rolf Willmann. 1982. *Economics of Artisanal and Mechanized Fisheries in Kerala: A Study on Costs and Earnings of Fishing Units*. Madras: FAO/UNDP.

Kurien, P.A. 1994. 'Non-economic Bases of Economic Behaviour: The Consumption, Investment and Exchange Patterns of Three Emigrant Communities in Kerala, India'. *Development and Change* 25: 757–783.

Kurup, P.G. 1972. 'Studies on the Physical Aspects of the Mud Banks along the Kerala Coast with special Reference to the Purakad Mud Bank'. *Bulletin of the Department of Marine Sciences*, University of Cochin 8: 1–72.

Lakoff, George. 1987. *Women, Fire and Dangerous Beasts: What Categories Reveal about the Mind.* Chicago: University of Chicago Press.

Lakoff, George and Mark Johnson. 1980. *Metaphors We Live By*. Chicago: University of Chicago Press.

Lave, Jean and Etienne Wenger. 1991. *Situated Learning: Legitimate Peripheral Participation*. Cambridge: Cambridge University Press.

Levinson, Steven. 2003. *Space in Language and Cognition: Explorations in Cognitive Diversity*. Cambridge: Cambridge University Press.

Lévi-Strauss, Claude. 1966. 'The Scope of Anthropology'. *Current Anthropology* 7: 112–123.

McGilvray, Dennis B. 1982. 'Sexual Power and Fertility in Sri Lanka: Batticaloa Tamils and Moors'. In: C.P. Cormack (ed.), *Ethnography of Fertility and Birth*. London: Academic Press, pp. 25–73.

MacPherson, H. and P.G. Kurup. 1981. 'Wave Damping at the Kerala Mud Banks'. *Indian Journal of Marine Sciences* 10: 154–160.

Mathur, P.R.G. 1978. *The Mappila Fisherfolk of Kerala*. Thiruvananthapuram: Kerala Historical Society.

Mathur, P.R.G. 1979. 'Hindu Marine Fisherfolk of Kerala – Some Ethnographic Notes'. *Journal of Kerala Studies* 6: 131–172.

Matsyafed. 1999. *An Overview of Activities*. Trivandrum: Kerala State Co-operative Federation for Fisheries Development.

Mauss, Marcel. 1954. *The Gift: Form and Function of Exchange in Archaic Societies*. Translated by Ian Cunnison. London: Routledge & Kegan Paul.

Menon, Dilip M. 1994. *Caste, Nationalism and Communism in South India: Malabar, 1900-1948*. Cambridge: Cambridge University Press.

Meynen, Wicky. 1989. 'Fisheries Development, Resources Depletion and Political Mobilization in Kerala: The Problem of Alternatives'. *Development and Change* 20: 735–770.

Miller, Roland E. 1992. *Mappila Muslims of Kerala: A Study in Islamic Trends*. 2nd edition. Madras: Orient Longman.

Moore, Melinda. 1990. 'The Kerala House as a Hindu Cosmos'. In: M. Marriott (ed.), *India Through Hindu Categories*. New Delhi: Sage, pp. 169–202.

Moreno, Manuel and McKim Marriott. 1990. 'Humoral Transactions in Two Tamil Cults: Murukan and Mariyamman'. In: M. Marriott (ed.), *India through Hindu Categories*. New Delhi: Sage, pp. 148–167.

Myerhoff, Barbara. 1992. *Remembered Lives: The Work of Ritual, Storytelling, and Growing Older*. Ann Arbor: University of Michigan Press.

Nair, R.V. 1952. 'Studies on the Revival of the Indian Oil Sardine Fishery'. In: *Proceedings of the Indo-Pacific Fisheries Council, 4th Meeting (Quezon City, Philippines)*, Section II, Bangkok: IPFC Secretariat, pp. 115–129.

Namputhirippattu, Kaanippayuur Shankaran. 1994. *Yanthravidhikal (bhaashaavya khyaanasahitham)* (in Malayalam). Kunnamkulam: Panchangam Publications.

Namputhirippaattu, Kaanippayuur Shankaran. 1997. *Manushyaalayachandrika* (in Malayalam). Kunnamkulam: Panchangam Publications.

Nandargikar, G.R. 1971. *The Raghuvamsha of Kaalidaasa with the Commentary of Mallinaatha*. Delhi: Motilal Banarsidass.

Nayak, Nalini. 1993. *Continuity and Change in Artisanal Fishing Communities: A Study of Socio-economic Conditions of Artisanal Fishing Communities on the South West Coast of India Following Motorisation of Fishing Crafts*. Thiruvananthapuram: Programme for Community Organisation.

Nichter, Mark. 1980. 'The Layperson's Perception of Medicine as Perspective into the Utilization of Multiple Therapy Systems in the Indian Context'. *Social Science and Medicine* 14B: 225–233.

Nichter, Mark. 1992. 'Of Ticks, Spirits and the Promise of Vaccines'. In: Charles Leslie (ed.), *Paths to Asian Medical Knowledge*. Berkeley: University of California Press, pp. 224–253.

Obeyesekere, Gananath. 1984. *The Cult of the Goddess Pattini*. Chicago: University of Chicago Press.

Okasha, Samir. 2002. *Philosophy of Science: A Very Short Introduction*. Oxford: Oxford University Press.

Osella, Caroline. 1993. 'Making Hierarchy Natural: The Cultural Construction of Gender and Maturity in Kerala, India'. Unpublished Ph.D. thesis, London.

Osella, Filippo and Caroline Osella. 1996. 'Articulation of Physical and Social Bodies in Kerala'. *Contributions to Indian Sociology* (NS) 30: 37–68.

Osella, Filippo and Caroline Osella. 1999. 'Seepage of Divinised Power through Social, Spiritual and Bodily Boundaries: Some Aspects of Possession in Kerala'. In: Jackie Assayag and Gilles Tarabout (eds), *La possession en Asie du Sud: Parole, Corps, Territoire, Purusharta 21. Paris: Éditions de l'École des Hautes Études en Sciences Sociales*, pp. 183–210.

Osella, Filippo and Caroline Osella. 2000. *Social Mobility in Kerala: Modernity and Identity in Conflict*. London: Pluto Press.

Padmanabha Menon, K.P. 1993 [1924]. *History of Kerala*, Vol. 3. New Delhi: Asian Educational Services.

Pálsson, Gísli. 1994. 'Enskilment at Sea'. *Man (N.S.)* 29:901–927.

Parayil, Govindan. 2000. 'Is Kerala's Development Experience a "Model"?' In: G. Parayil (ed.). *Kerala: The De*velopment Experience: Reflections on Sustainability and Replicability. London: Zed Publications, pp. 1–15.

Parpola, Asko. 1994. *Deciphering the Indus Script*. Cambridge: Cambridge University Press.

Parpola, Asko. 1997. 'The Iconography and Cult of Kuttichattan: Field Research on the Sanskritization of Local Folk Deities in Kerala'. In: Johannes Bronkhorst, and Madhav M. Deshpande (eds), *Aryan and Non-Aryan in South Asia: Evidence, Interpretation and Ideology*. Cambridge, Mass.: Harvard University Press, pp. 175–205.

PCO/SIFFS. 1991a. *Techno-Economic Analysis of Motorisation of Fishing Units Along the Lower South-West Coast of India*. Thiruvananthapuram:

Programme for Community Organisation/South Indian Federation of Fishermen Societies.

PCO/SIFFS. 1991b. *Motorisation of Fishing Units: Benefits and Burdens.* Thiruvananthapuram: Programme for Community Organisation/ South Indian Federation of Fishermen Societies.

Penzer, N.M. 1968 [1924]. *The Ocean of Stories, being C.H. Towney's translation of Somadeva's Katha Sarit Sangara (or Ocean of Streams of Story).* Delhi: Motilal Banarsidass.

Pillai, T.S. 1962. *Chemmeen* [*Cemmīn*]. Translated by Narayana Menon. Mumbai: Jaico.

Piot, Charles. 1999. *Remotely Global: Village Modernity in Western Africa.* Chicago: University of Chicago Press.

Platteau, J. 1984. 'The Drive Towards Mechanization of Small-scale Fisheries in Kerala: A Study of the Transformation Process of Traditional Village Societies'. *Development and Change* 15: 65–103.

Prakash, Gyan. 1999. *Another Reason: Science and the Imagination of Modern India.* Princeton: Princeton University Press.

Purushooththaman, Paayakkattu. 1995. *Paalappetti ksheethracharithram* (in Malayalam), in: *Shrii paalapetti bhagavathi ksheethram naviikarana kalasham.* Palappetti.

Radcliffe-Brown, A.R. 1952. *Structure and Function in Primitive Society.* London: Cohen & West.

Rajan, J.B. 1993. *Techno-Socio-Economic Study on Ringseine Fishing in Kerala.* Thiruvananthapuram: Programme for Community Organisation.

Ram, Kapila. 1991. *Mukkuvar Women: Gender, Hegemony and Capitalist Transformation in a South Indian Fishing Community.* London: Zed Publications.

Ramachandran Nair, K.K. 1986. *Kerala State Gazetteer,* Vol. I. Thiruvananthapuram: Government of Kerala.

Ramanujan, A.K. 1967. *The Interior Landscape: Love Poems from a Tamil Anthology.* Bloomington: Indiana University Press.

Rao Bahadur Govindan Memorial Committee. 1973. *Souvenir.* Kaipamangalam.

Richards, Paul. 1993. 'Cultivation: Knowledge or Performance?' In: Mark Hobart (ed.), *An Anthropological Critique of Development: The Growth of Ignorance.* London: Routledge, pp. 61–78.

Ricoeur, Paul. 1985. *Die lebendige Metapher.* Translated into German by Rainer Rochlitz, Munich: Wilhelm Fink Verlag.

Rorty, Richard. 1987. 'Hesse and Davidson on Metaphor'. *The Aristotelian Society,* Supplement 61: 283–296.

Sahlins, Marshall. 1972. *Stone Age Economics.* Chicago: Aldine.

Schimmel, Annemarie. 1975. *Mystical Dimensions of Islam.* Chapel Hill: University of North Carolina Press.

Scott, Colin. 1989. 'Knowledge Construction among Cree Hunters: Metaphors and Literal Understanding'. *Journal de la Société des Américanistes* 75: 193–208.

Scott, James C. 1990. *Domination and the Arts of Resistance: Hidden Transcripts*. New Haven: Yale University Press.

Scott, James C. 1998. *Seeing Like a State: How Certain Schemes to Improve the Human Condition Have Failed.* New Haven: Yale University Press.

Séhouéto, Lazare M. 1995. '"Lokales" Wissen und bäuerliche Naturaneignung in Benin'. In: Ute Luig and A. von Oppen (eds.), *Naturaneignung in Afrika als sozialer und symbolischer Prozeß*. Berlin: Das Arabische Buch, pp. 83–94.

Shankunni, Kottaaraththil. 1998 [1909–1934]. *Eithihyamaala: ettu bhaagangalum* (in Malayalam). Kottayam: DC Books.

SIFFS. 1992. *A Census of the Artisanal Marine Fishing Fleet of Kerala 1991.* Thiruvananthapuram: South Indian Federation of Fisheries Societies.

SIFFS. 1999. *A Census of the Artisanal Marine Fishing Fleet of Kerala 1998.* Thiruvananthapuram: South Indian Federation of Fisheries Societies.

Silas, E.G. (ed.). 1984. 'Mudbanks of Kerala Coast'. *Central Marine Fisheries Research Institute Bulletin* 31.

Söhnen-Thieme, Renate. 1995. 'On the Concept and Function of sathya ("truth") in Ancient Indian Literature'. In: Cezary Galewicz (ed.), *Proceedings of the International Conference on Sanskrit and Related Studies. Cracow Indological Studies, vol. I.* Cracow: Enigma Press, pp. 235–244.

Sperber, Dan and Deirdre Wilson. 1995. *Relevance: Communication and Cognition.* 2nd edition. Oxford: Blackwell.

Sunish, C.D. 1995. 'Velumukkuvan samsaarikkunu' (in Malayalam). In: *Keeraliyathayute naattarivu* 2: 106–110.

Suseelan, C., K.N. Rajan and G. Nandakumar. 1989. 'The "Karikkadi" Fishery of Kerala'. *Marine Fisheries Extension Series, Technical and Extension Series* 102: 4–8.

Tarabout, Gilles. 1986. *Sacrifier et donner a voir: Les temples de Kerala, Inde du Sud, étude anthropologique.* Paris: École Française d'Extreme Orient.

Tarabout, Gilles. 1993. 'Violence et Non-violence Magiques. La Sorcellerie au Kérala'. *Purusartha* 16: 155–185.

Tarabout, Gilles. 1997a. 'Maîtres et serviteurs. Commander à des dieux au Kerala (Inde du Sud)'. In: Albert de Surgy, André Padoux and Pierre Lory (eds), *Religion et pratiques de Puissances*. Paris: L'Harmattan, pp. 253–284.

Tarabout, Gilles. 1997b. 'Une Configuration Sociale Régionale: le Kérala'. *Historiens et Géographes* 356: 243–256.

Tarabout, Gilles. 1999. '"Psycho-Religious Therapy" in Kerala as a Form of Interaction between Local Traditions and (perceived) Scientific Discourse'. In: Marine Carrin (ed.), *Managing Distress: Possession and Therapeutic Cults in South Asia*. New Delhi: Manohar, pp. 133–154.

Thurston, Edgar. 1909. *Castes and Tribes of South India*, 7 volumes. Madras: Government Press.

Tsing, Anna Lowenhaupt. 1993. *In the Realm of the Diamond Queen: Marginality in an Out-of-the-way Place*. Princeton: Princeton University Press.

Tsing, Anna Lowenhaupt. 2005. *Friction: An Ethnography of Global Connection*. Princeton: Princeton University Press.

Turner, Howard R. 1995. *Science in Medieval Islam: An Illustrated Introduction*. Austin: University of Texas Press.

Uchiyamada, Yasushi. 1995. 'Sacred Grove (Kaavu): Ancestral Land of "Landless Agricultural Labourers" in Kerala, India'. Unpublished Ph.D. thesis, London.

Udayabhanu, K.M. 1990. 'The Diivara Community and Social Change in Kerala'. Unpublished Ph.D. thesis, Thiruvananthapuram.

Vaidyanathan, K.R. 1982. *Temples and Legends of Kerala*. Bombay: Bharatiya Vidya Bhavan.

Valeri, Valerio. 2000. *The Forest of Taboos: Morality, Hunting, and Identity among the Huaulu of the Moluccas*. Madison: University of Wisconsin Press.

Vasavi, A.R. 1999. *Harbingers of Rain: Land and Life in South India*. New Delhi: Oxford University Press.

van der Veer, Peter. 1994. *Religious Nationalism: Hindus and Muslims in India*. Berkeley: University of California Press.

Velukutti Arayan, V.V. 1956. *Thakazhayute Chemmiin: Oru Nirupanam ("Thakazhi's Chemmiin: A Critical Review")* (in Malayalam). Thiruvananthapuram: Kalakeralam Publications.

Veyne, Paul. 1988. *Did the Greeks Believe in their Myths? An Essay on the Constitutive Imagination*. Translated by Paula Wissing. Chicago: University of Chicago Press.

Vico, Giambattista. 1968 [1744]. *The New Science of Giambattista Vico*. Translated by Thomas Goddard Bergin and Max Harold Fisch. Ithaca and London: Cornell University Press.

Vishalakshmi, N.R. 1973. 'Arayanmar' (in Malayalam). In: *Kerala charithram. Ernakulam:* Kerala History Association, vol. 1: pp. 992–1013.

Wagner, Roy. 1977. 'Scientific and Indigenous Papuan Conceptualizations of the Innate: A Semiotic Critique of the Ecological Perspective'. In: T. Bayliss-Smith and R. Feacham (eds), *Subsistence and Survival*. London: Academic Press, pp. 385–410.

Wagner, Roy. 1981. *The Invention of Culture*. Chicago: University of Chicago Press.

Wallich, Paul. 1995. 'A Mystery Inside a Riddle Inside an Enigma'. *Scientific American* 272, March, 19.

Warrier, M.I., E.P. Narayana Bhattathiri and K.R. Warrier. 1999. *Malayaalam-Imglish Nighantu (Malayalam–English Dictionary).* Kottayam: D.C. Books.

Weber, Max. 1920 [1905]. 'Die protestantische Ethik und der Geist des Kapitalismus'. In: *Gesammelte Aufsätze zur Religionssoziologie,* Vol. I. Tübingen: Mohr.

White, Hayden. 1973. *Metahistory: The Historical Imagination in Nineteenth-Century Europe.* Baltimore: Johns Hopkins University Press.

Whitehead, Harvey (ed.). 2001. *The Debated Mind: Evolutionary Psychology versus Ethnography.* Oxford and New York: Berg.

Wilson, James A. 1990. 'Fishing for Knowledge'. *Land Economics* 66: 12–29.

Yohannan, T.M., P.N. Radhakrishnan, N.G.K. Pillai and P.L. Ammini. 2000. 'Marine Fisheries in Kerala'. Mimeo draft as appendix for Balakrishnan II report.

Ziener, Paul B. and Kjeld Rasmussen. 1958. *Report No. 1 to the Government of India on Fishing Boats.* Rome: FAO.

Zimmermann, Francis. 1983. 'Remarks on the Conception of the Body in Ayurvedic Medicine'. *South Asian Digest of Regional Writing* 8: 10–25.

Zimmermann, Francis. 1987. *The Jungle and the Aroma of Meats: An Ecological Theme in Hindu Medicine.* Translated by Janet Lloyd. Berkeley: University of California Press.

Zimmermann, Francis. 2003. 'Bodily Humors in the Scholarly Tradition of Hindu and Galenic Medicine as an Example of Naive Theory and Implicate Universals'. In: Glauco Sanga and Gherardo Ortalli (eds), *Nature Knowledge: Ethnoscience, Cognition and Utility.* New York and Oxford: Berghahn Books, pp. 262–271.

Index

www.ingramcontent.com/pod-product-compliance
Ingram Content Group UK Ltd.
Pitfield, Milton Keynes, MK11 3LW, UK
UKHW021838100926
8556IPUK00004B/408

9 781845 450151